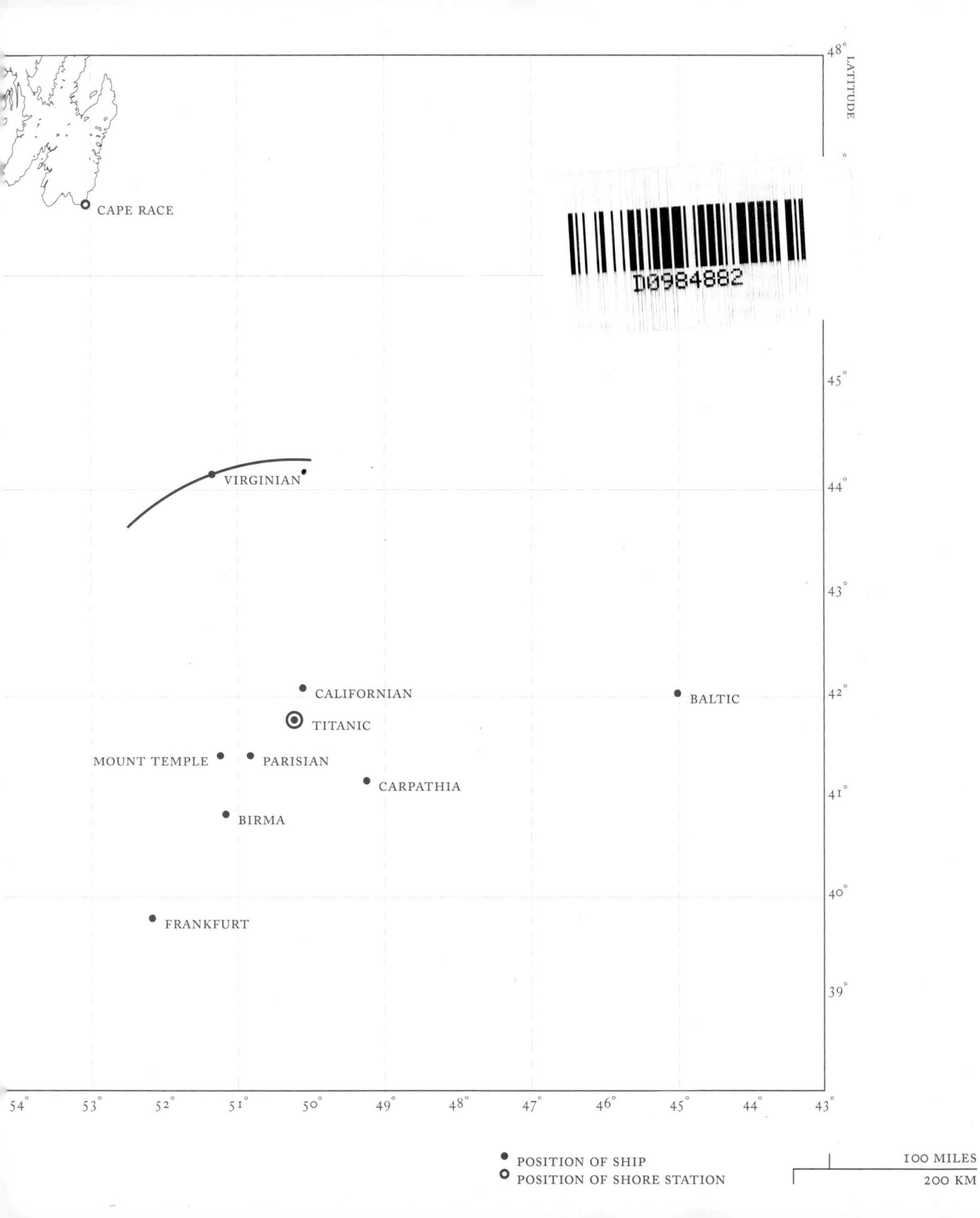
48°
LATITUDE
CAPE RACE
D0984882
45°
VIRGINIAN
44°
43°
CALIFORNIAN
BALTIC
42°
TITANIC
MOUNT TEMPLE
PARISIAN
CARPATHIA
41°
BIRMA
40°
FRANKFURT
39°
54°
53°
52°
51°
50°
49°
48°
47°
46°
45°
44°
43°
POSITION OF SHIP
POSITION OF SHORE STATION
100 MILES
200 KM

TITANIC CALLING

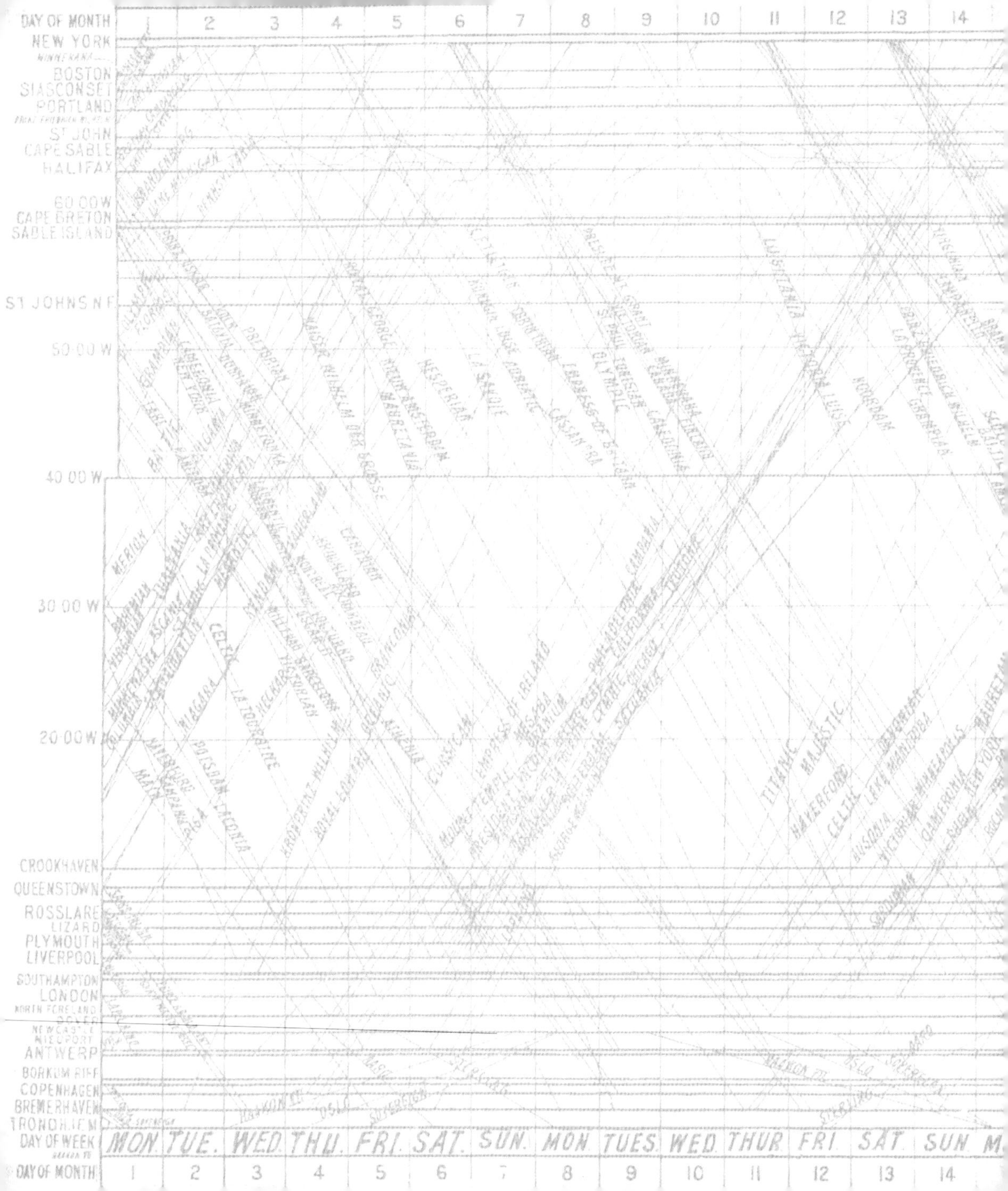
DAY OF MONTH
1
2
3
4
5
6
7
8
9
10
11
12
13
14
NEW YORK
BOSTON
SIASCONSET
PORTLAND
ST JOHN
CAPE SABLE
HALIFAX
60 00 W
CAPE BRETON
SABLE ISLAND
ST JOHNS N F
50 00 W
40 00 W
30 00 W
20 00 W
CROOKHAVEN
QUEENSTOWN
ROSSLARE
LIZARD
PLYMOUTH
LIVERPOOL
SOUTHAMPTON
LONDON
NORTH FORELAND
DOVER
NEWCASTLE
ANTWERP
BORKUM RIFF
COPENHAGEN
BREMERHAVEN
TRONDHJEM
DAY OF WEEK
MON.
TUE.
WED.
THU.
FRI.
SAT.
SUN.
MON
TUES.
WED.
THUR
FRI
SAT.
SUN
LUSITANIA
OLYMPIC
HESPERIAN
LA SAVOIE
MERION
CELTIC
NIAGARA
LA TOURAINE
CORSICAN
EMPRESS OF IRELAND
MESABA
TITANIC
MAJESTIC
HAVERFORD
CELTIC
LAKE MANITOBA
AUSONIA
CAMERONIA
NEW YORK
MAURETANIA
OSLO
HAAKON VII
SOMERSET
STAVANGERFJORD
NOORDAM
CARONIA
CAMPANIA
TEUTONIC
CHICAGO
OCEANIC
VICTORIAN
CARMANIA
CALEDONIA
CARPATHIA
SAXONIA

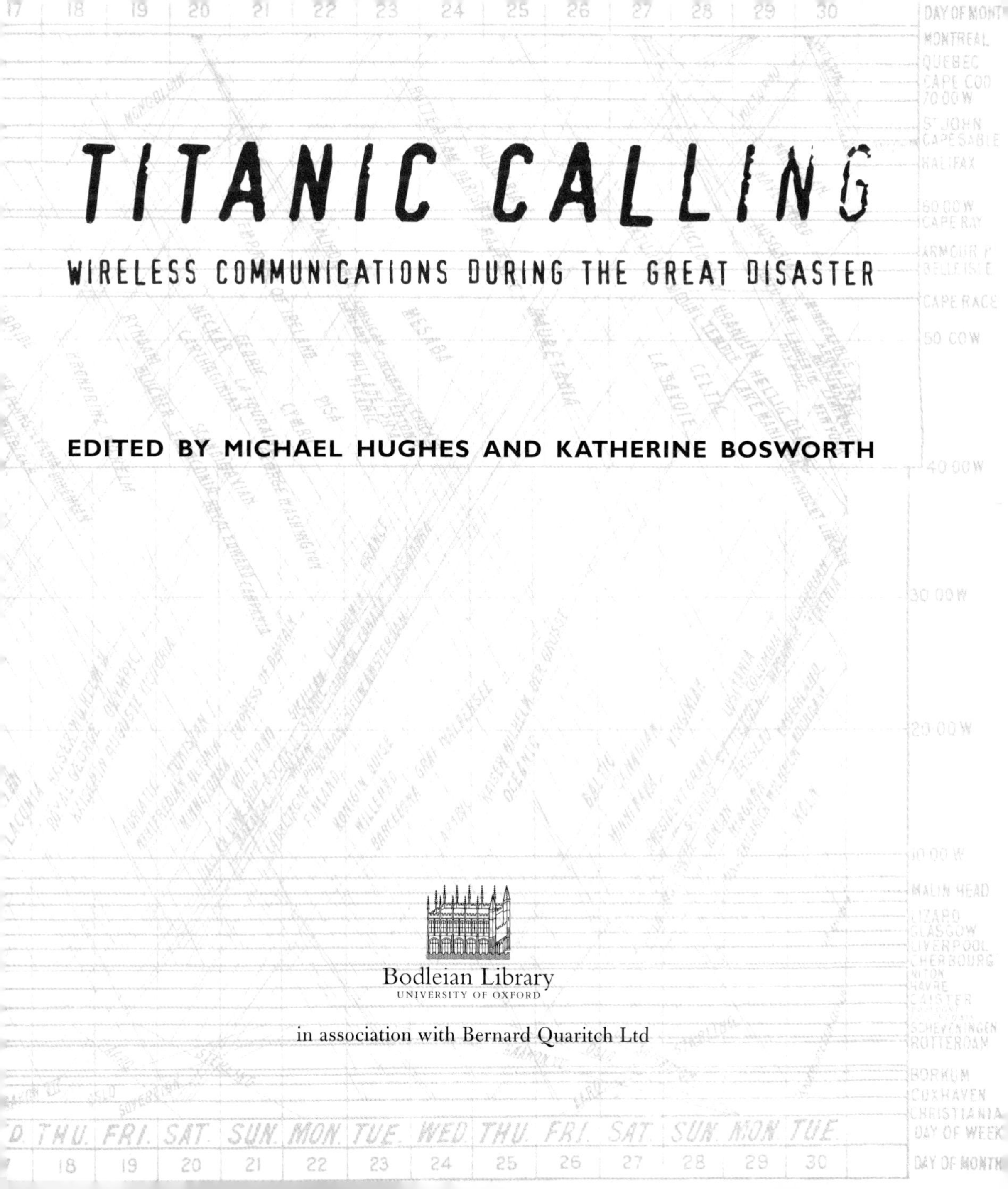

TITANIC CALLING

WIRELESS COMMUNICATIONS DURING THE GREAT DISASTER

EDITED BY MICHAEL HUGHES AND KATHERINE BOSWORTH

Bodleian Library
UNIVERSITY OF OXFORD

in association with Bernard Quaritch Ltd

First published in 2012 by the Bodleian Library
Broad Street Oxford OX1 3BG

www.bodleianbookshop.co.uk

in association with Bernard Quaritch Ltd

ISBN 978 1 85124 377 8

Cover design by Dot Little
Designed and typeset by illuminati, Grosmont
in 11 on 13 Monotype Bulmer
Endpaper maps designed by Joby Ellis
Printed and bound by Great Wall Printing Company Ltd, Hong Kong,
on 157 gsm Go Ching matt art

British Library Catalogue in Publishing Data
A CIP record of this publication is available from the British Library

CONTENTS

ACKNOWLEDGEMENTS

We are grateful to John Koh for the ideas that led to the conception of this book. Particular thanks are due to Alexander Whiscombe for his assistance with the transcription, and to Joby Ellis and Michael Mack for their work in designing the maps. Thanks also to the staff of the Special Collections Reading Room of the Bodleian Library for their unfailing help and patience, and especially to Julia Wagner for her help with translation. John Sayers kindly provided images of the *Titanic* from his private collection, and we are grateful to the Museum of the History of Science for permission to use images. We would also like to thank Samuel Fanous, Deborah Susman and other staff of Bodleian Library Publishing for all their help.

NOTE ON TRANSCRIPTS

During the process of transcribing the original material in this book, certain amendments became necessary in order to highlight and make accessible the important and compelling details of the communications. With this in mind, formatting of the procès-verbaux and messages has been standardised and some punctuation (particularly capitalisation) has been added. On the PVs, a column showing New York Time (NYT) has been added for all ships and shore stations which were not already working in this time zone, in order to facilitate comparison between them. It should be noted that the conversion to New York Time is an estimate, based on information contained within the messages, and from the statements of wireless operators given before the official inquiries into the disaster. Columns (often blank) for notes on the quality and duration of the signals and the operator's initials have been removed for simplicity. Some entries have also been omitted (indicated on the transcript) where there were large sections which were not related to the *Titanic* (indicated on the transcript by ellipses). In the individual messages, information relating to the operator handling the message and any retransmission of the message have been removed, leaving as prefatory material only the sender, recipient, date and the office at which the message was recorded. The times of messages have also been omitted, owing to the difficulty of establishing the original time sent, given the inevitable delays involved in (sometimes multiple) relaying of messages. In as far as possible, however, messages have been placed in their correct chronological sequence. The only exception to this is the messages to and from Bruce Ismay in the aftermath of the sinking, where times have been included in order to illustrate the rapidity of the dialogue.

It is hoped that these minor changes will not detract from the tension and immediacy conveyed by the messages, but will enable a wider appreciation of the material, some of which is published here for the first time.

PREFACE

In December 2004 the extensive Marconi Collection was generously given to the University of Oxford, successfully concluding the search for a permanent and secure home for both the archives and the historic equipment accumulated by Marconi plc and its predecessor companies. The Marconi Archives are held in the Bodleian Library, while the equipment is now housed in the adjacent Museum of the History of Science.

Records relating to the *Titanic* disaster form a series of especial significance within the Marconi Archives. Gathered together by the Marconi International Marine Communication Company in its endeavours to provide to the British Inquiry all possible documentation with regard to communications by wireless telegraphy during the disaster, they form a unique record of the events that took place on the night of 14/15 April 1912. In the centenary year of the sinking, this volume draws upon this compelling resource to tell the story of the voyage and its catastrophic ending as revealed by the wireless communications. The narrative is complemented by an extensive selection of transcripts of logs and messages that document events as they occurred.

The Bodleian Library is very grateful for the support of Bernard Quaritch Ltd in the production and publication of this book, which has been a collaborative effort between the two organisations. Quaritch's interest in the *Titanic* was first inspired by a telegram in the company's archive: a brief message from the American bookseller A.S.W. Rosenbach to B.A. Quaritch (son of the company's founder) to notify him of the loss of Harry Elkins Widener in the sinking. Widener, son of the wealthy businessman George D. Widener, was returning to New York with his parents after a visit to London and Paris. A keen book collector, Widener had first been introduced to Quaritch by Rosenbach in 1907 and over the following five years became one of the firm's most enthusiastic customers. Whilst in London in 1912 Widener had frequently visited Quaritch, making purchases, placing bids (to be executed by Quaritch) for a number of lots in the forthcoming Huth sale, Part II (June 1912), and arranging for shipment of his purchases. Although the majority of his books were sent together, a 1598 edition of Francis Bacon's *Essaies*, purchased the previous November in the first Huth sale, is marked on Quaritch's invoice as simply 'delivered'. Both Quaritch and Rosenbach told the story, after the sinking, that Widener had refused to be parted from this book, taking it with him on the *Titanic* and (according to his mother, who survived) placing it in his pocket as the ship went down. The news of Widener's death, when received from Rosenbach, must have been a significant blow to Quaritch but it did not quite mark the end of Widener's involvement with either bookseller. Quaritch went on to purchase eighteen lots on Widener's behalf in the second Huth sale (the bids commissioned prior to his death) and Mrs Widener continued to build her son's collection, making numerous purchases with Rosenbach and several with Quaritch, before founding the Harry Elkins Widener Memorial Library at Harvard University, which opened in 1915. This publication is, in a small way, a tribute to one of Quaritch's finest customers, one hundred years after his premature death.

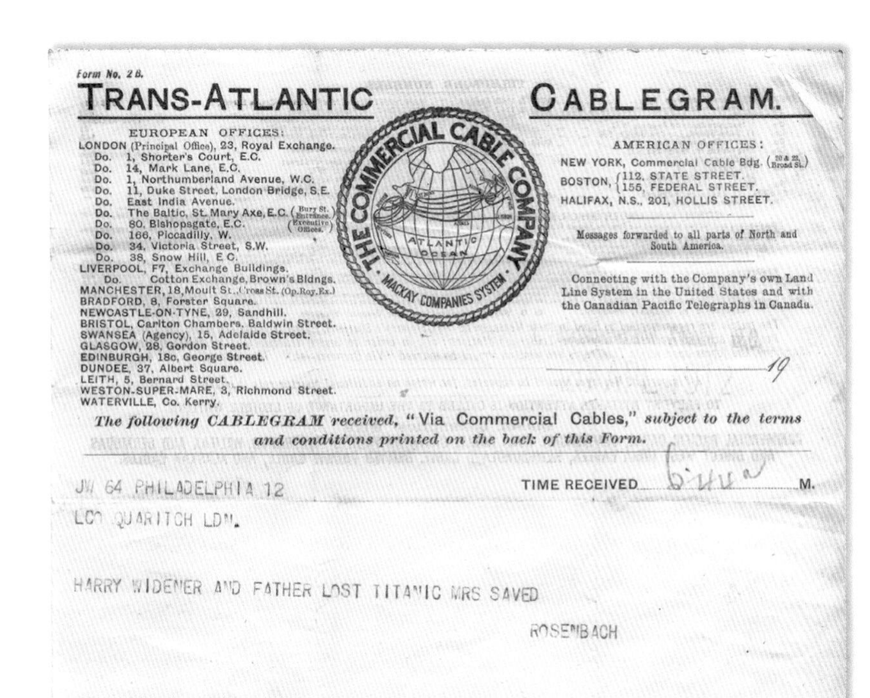

Form No. 2B.

TRANS-ATLANTIC CABLEGRAM.

EUROPEAN OFFICES:
LONDON (Principal Office), 23, Royal Exchange.
Do. 1, Shorter's Court, E.C.
Do. 14, Mark Lane, E.C.
Do. 1, Northumberland Avenue, W.C.
Do. 11, Duke Street, London Bridge, S.E.
Do. East India Avenue.
Do. The Baltic, St. Mary Axe, E.C. (Bury St. Entrance.)
Do. 80, Bishopsgate, E.C. (Executive Offices.)
Do. 166, Piccadilly, W.
Do. 34, Victoria Street, S.W.
Do. 38, Snow Hill, E.C.
LIVERPOOL, F7, Exchange Buildings.
Do. Cotton Exchange, Brown's Bldngs.
MANCHESTER, 18, Moult St., Cross St. (Op. Roy. Ex.)
BRADFORD, 8, Forster Square.
NEWCASTLE-ON-TYNE, 29, Sandhill.
BRISTOL, Carlton Chambers, Baldwin Street.
SWANSEA (Agency), 15, Adelaide Street.
GLASGOW, 28, Gordon Street.
EDINBURGH, 18c, George Street.
DUNDEE, 37, Albert Square.
LEITH, 5, Bernard Street.
WESTON-SUPER-MARE, 3, Richmond Street.
WATERVILLE, Co. Kerry.

THE COMMERCIAL CABLE COMPANY · MACKAY COMPANIES SYSTEM · ATLANTIC OCEAN

AMERICAN OFFICES:
NEW YORK, Commercial Cable Bdg. (20 & 22 Broad St.)
BOSTON, 112, STATE STREET. 155, FEDERAL STREET.
HALIFAX, N.S., 201, HOLLIS STREET.

Messages forwarded to all parts of North and South America.

Connecting with the Company's own Land Line System in the United States and with the Canadian Pacific Telegraphs in Canada.

19

The following CABLEGRAM received, "Via Commercial Cables," *subject to the terms and conditions printed on the back of this Form.*

JW 64 PHILADELPHIA 12 TIME RECEIVED 6.44 a M.

LCO QUARITCH LDN.

HARRY WIDENER AND FATHER LOST TITANIC MRS SAVED

ROSENBACH

Telegram from the American bookseller A.S.W. Rosenbach to Quaritch, informing him of the death of Harry E. Widener. (Bernard Quaritch Ltd.)

The last photograph of the *Titanic*, taken as she left Queenstown, Ireland, 11 April.

WIRELESS TELEGRAPHY AND THE *TITANIC* DISASTER

THE ARCHIVAL RECORD

Documentary material in the Marconi Archives relating to the sinking of the RMS *Titanic* owes its survival to the inquiries that took place in the months following the sinking. The collection contains almost no similar material for other events, great or small: it would appear that the records of wireless stations (both ship and shore) were not otherwise systematically retained. For the days and weeks surrounding the *Titanic*'s fateful voyage there does exist a series of procès-verbaux (radio operators' logs) for ships in the vicinity of the *Titanic* at the time, together with the station records of the messages sent and received, and other data relating to the event.

The records are incomplete – parts either do not survive, or appear to have found their way elsewhere – but the documentary record is probably the most extensive, and provides the most direct written evidence of the events that took place, of any to survive. The only parallel is the body of evidence submitted by passengers and crew at the US and British inquiries into the sinking, and copies of these are to be found in the collection too. It is thanks to the use of wireless telegraphy that there were any survivors at all, underlining the significance of the record, though it is important to place in context the rescue of 712 people against the loss of some 1,517 souls in the disaster. The process of gathering the necessary evidence is itself preserved in the Archives, in the form of correspondence with other companies employing the radio operators involved, seeking the relevant documents. Sometimes the original documents were sent; in other cases the company concerned transcribed the relevant procès-verbaux afresh. Service forms recording the messages handled by the North American shore stations were also accumulated, creating modest series in several instances, but those for the Camperdown station run into thousands, reflecting the significance of this station in handling much of the RMS *Carpathia*'s wireless traffic in the days following the sinking.

This account aims to document the sinking using those sources most pertinent as a record of events. Attention is focused on several aspects: a summary of the development of wireless telegraphy up to 1912; the training and role of the wireless operators; the wireless record of the calamity as it happened; the use of the documentation in the British Inquiry to establish the sequence of events before the disaster – especially with regard to ice warning messages between ships crossing the Atlantic in the days before the disaster; and the information about the role of wireless telegraphy to be gleaned from the American and British inquiries into the circumstances of the sinking. Documents and messages are

3722.A.
ROTARY PHOTO E.C.
WHITE STAR LINER TITANIC.
LENGTH 882 ft. 6 ins. BREADTH 92 ft. 6 ins 45,000 TONNAGE.
SAILED FROM SOUTHAMPTON ON HER ILL-FATED MAIDEN VOYAGE ON APRIL 10TH, 1912, CARRYING 2,350 PASSENGERS AND CREW. STRUCK AN ICEBERG
OFF THE COAST OF NEWFOUNDLAND. PERISHED ON SUNDAY NIGHT APRIL 14TH, 1912.

The White Star Liner *Titanic*, 1912. The wireless aerial is just visible. (The Sayers Collection.)

used to illuminate events, and a number of these are included in full in the transcripts forming the second part of this volume. The sets of transcripts provide additional detail supplementing the account given here.

SETTING THE SCENE: THE DEVELOPMENT OF MARITIME WIRELESS TELEGRAPHY TO 1912

An article on the career prospects for a wireless operator in 1911 commented that before Marconi's 'invention', communications between ships and from ship to shore were possible only over short distances by flag signals during daylight and Morse signal lamp at night. 'Now, by means of the Marconi system, communication can be maintained over hundreds of miles.'

Publicity booklet produced by Marconi's Wireless Telegraph Company, *c.* 1910. (Oxford, Bodleian Library, MS. Marconi 1705, *Marconi Wireless Telegraphy: A Short History*, cover.)

The development of wireless telegraphy created a new type of telegraphist known as the 'Marconi operator'. There were several hundred of these, with a growing demand for skilled men.[1]

Marconi's Wireless Telegraph Company had been set up in 1897 (originally as the Wireless Telegraph and Signal Company), with backing from Guglielmo Marconi's Anglo-Irish cousin Henry Jameson Davis and his business associates. Initially the company focused on experimental work to develop Marconi's patented equipment, to make it effective in practical applications especially in the maritime field. Marconi himself, born in 1874 to Giuseppe Marconi and his wife Annie (née Jameson), had come to Britain in 1896 to seek backing to develop and apply his research into wireless telegraphy. While others were working on wireless technology at the same time and no one individual can properly be said to have 'invented' wireless telegraphy, in these early days Marconi was the most successful in applying his own apparatus practically and commercially.

By the turn of the twentieth century, work was progressing on improvement of the techniques of wireless telegraphy. One significant development was the introduction of tuning. The effect of tuning was discovered by Oliver Lodge in 1897; Marconi's success in this sphere lay in applying the principle to his wireless equipment to improve its efficiency and create the ability to distinguish between messages sent at different wavelengths. This opened up the practical possibilities enormously, as it enabled more than two stations to communicate simultaneously without interference. Marconi's work was encapsulated in his well-known patent for tuning, no. 7777, in the year 1900. The use of the principle of tuning by Marconi led to dispute with Lodge over possible breach of his existing patents. The conflict was eventually resolved with the purchase of this and other Lodge patents by Marconi's company.

From the first, a significant potential use of wireless telegraphy lay in maritime communication. This market was exploited to the full by Marconi, and impelled the formation of a subsidiary company, the Marconi International Marine Communication Company (hereafter referred to as the Marine Company), in 1900.

A German liner, the SS *Kaiser Wilhelm der Grosse*, was the first merchant ship to be fitted with Marconi equipment, in February 1900. The first use of Marconi equipment on a British merchant ship was on the SS *Lake Champlain* in May 1900. The earliest surviving 'Marconigram' – a telegraph message sent using an official form produced by one of Marconi's companies – is a message from Marconi's Wireless Telegraph Company in Crookhaven, Ireland, to the operator on the *Lake Champlain*, sent in June 1901. By 1902, General Orders of the Marconi International Marine Communication Company had been produced. These detailed the procedures to be followed by radio operators on board ships, and were regularly updated. The operators were employed by the Company and their services contracted to the ship in question together with the

opposite
Guglielmo Marconi on board the *Lucania*, 1903. (Oxford, Bodleian Library, MS. Photogr. b. 60, fol. 4.)

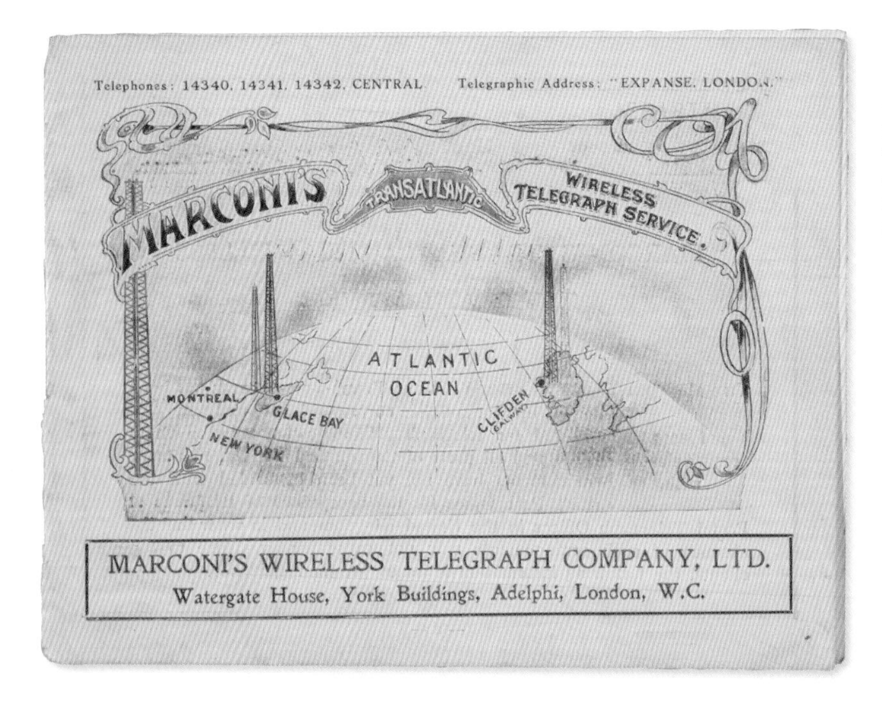

Publicity brochure for Marconi's Transatlantic Wireless Telegraph Service, c. 1907. (Oxford, Bodleian Library, MS. Marconi 218, cover.)

wireless equipment installed on board. In this early period, interest was also kindled in such organisations as Lloyd's of London, the major shipping insurance firm, which was alert to the possibilities of wireless for communicating with remote points for news of ships. A contract to equip ten of Lloyd's signal stations was signed in 1901.

In 1901, Marconi persuaded his company to invest heavily in attempting to send a signal from one side of the Atlantic to the other, to demonstrate the potential of the medium. At this time success was by no means assured. Opinion in the scientific community tended to the view that sending a signal such a huge distance was not possible because of the curvature of the earth – electromagnetic waves were known to travel in straight lines and it was not thought feasible to achieve transmission through the earth itself. Marconi (perhaps supported by the success of experiments in which signals were successfully transmitted progressively further distances regardless of intervening obstacles) was convinced otherwise and was prepared to put it to the test.

High-power wireless stations were constructed at Poldhu in Cornwall and Cape Cod on the US coast to facilitate this, but the aerials at both locations – supported by rings of masts – succumbed to gales and were destroyed almost as soon as they were put up. Undaunted, Marconi arranged for the construction of a temporary aerial at Poldhu and travelled with his assistants Kemp and Paget to an alternative location across the Atlantic: Signal Hill near St John's, Newfoundland. There, in December 1901, he set up his equipment and prepared for the reception of a prearranged signal of three dots (the letter S in Morse) to be transmitted at stipulated times from Poldhu. At St John's it was decided to use a balloon- or kite-supported aerial for receipt of the signal. These arrangements appear remarkably makeshift given the considerable investment in the original proposals: a measure perhaps of the sense of urgency felt by Marconi to achieve this landmark transmission.

Attempts were made to raise the aerial using balloons, but problems were encountered owing to high winds. Use of the kites proved more reliable, though one of these was lost too. On 12 December they finally heard the three dots repeated on a telephone in series with a sensitive detector, though the signal was not strong enough to register on the inker. Marconi's announcement of success was hailed by many, and resulted in a dinner held on 13 January 1902 in Marconi's honour by the American Institute of Electrical Engineers.

However, while Marconi and Kemp appeared confident that they heard the test signals in Newfoundland, there were some who doubted whether the signal had truly been heard, and Marconi was anxious to provide absolute proof of the capability of his equipment. On a voyage to New York on the SS *Philadelphia* in February 1902, two months after the transatlantic tests, he successfully received signals at increasingly greater distances from Poldhu – the source of the transmission – and a track chart illustrating these signals was certified by the master of the ship. These signals were the conclusive proof that transatlantic communication was possible.

Further key advances in the first decade of the twentieth century included first the development by Marconi of a magnetic receiver to replace the coherer used to detect signals in the earliest equipment, and then the invention by Dr J.A. Fleming (later Sir Ambrose), a leading figure in research into wireless telegraphy and for many years scientific adviser to Marconi's Wireless Telegraph Company, of the thermionic valve or diode. The valve rectified electrical oscillations so that they could be detected more easily, thus greatly facilitating the interpretation of wireless transmissions. Further development of the valve, by others as well as Marconi and Fleming, ultimately led to the transmission of speech by wireless, and the birth of sound broadcasting.

One perhaps unexpected phenomenon observed during this period was the effect of daylight on wireless transmission: reception was considerably reduced compared to night-time communication. This phenomenon was known but not fully understood for some years, and was due to bending of electromagnetic waves by an ionised layer in the atmosphere whose properties were altered by the sun's rays. It was this reflection that enabled long-distance wireless transmission to take place. As maritime services became more sophisticated, this

Cover for a Canadian Pacific ocean newspaper, 1916. (Oxford, Bodleian Library, MS. Marconi 252, *Canadian Pacific Empress Mail*, cover.)

THE CANADIAN PACIFIC OCEAN SERVICES LTD.
Managers and Agents for:—
CANADIAN PACIFIC RAILWAY OCEAN STEAMSHIP LINES, & THE ALLAN LINE STEAMSHIP COMPANY LTD.

MARCONIGRAMS

CANADIAN PACIFIC EMPRESS MAIL

left
Centre pages from Canadian Pacific ocean newspaper giving news Marconigrams, 1916. (Oxford, Bodleian Library, MS. Marconi 252, *Canadian Pacific Empress Mail*, centre pages.)

effect was exploited by making transmissions at night where possible, for example in the provision of ocean newspapers. These originated in one produced on the SS *St Paul* as early as 1899, consisting of a single sheet of news for the benefit of passengers on the liner as it neared Britain on a voyage from the United States. The news was relayed from the Needles station to Guglielmo Marconi on board the ship. With improvements in the range of communication, news was conveyed to ships at more distant points on their voyages, and incorporated into preprinted newspapers containing more general articles as well as space to add the news bulletins.

opposite
The Ocean Wireless News, *c.* 1915. (Oxford, Bodleian Library, MS. Marconi 252, *Ocean Wireless News*, cover.)

Ocean Wireless News

Ten Cents The Copy

Westward Ho! Number

THE HONEYMOON

THE
WIRELESS
AGE
APRIL
Third Article
on
HOW TO CONDUCT
A RADIO CLUB
In this Issue
FIFTEEN CENTS

The periodical *The Wireless Age* for April 1914. (Oxford, Bodleian Library, MS. Marconi 1113, *The Wireless Age*, cover.)

From these beginnings, maritime applications of wireless technology for communication developed steadily and had been in active use for more than a decade by 1912. Wireless equipment was installed on a number of vessels, their numbers increasing with each annual report produced by the Marine Company. In addition, associated companies overseas provided similar services using the Marconi system. Between them these had successfully contracted with several of the major transatlantic shipping lines to provide wireless equipment on board their vessels.

THE WHITE STAR CONTRACT AND THE *TITANIC*'S WIRELESS EQUIPMENT

The contract for provision of wireless telegraphy services to the White Star Line was similar to other agreements entered into by the Marine Company. The agreement of 1909 (extended in 1911 to include the *Titanic* and RMS *Olympic*) stipulated that the Marine company would provide and maintain the equipment and pay the operators' salaries, while the shipowners would provide a wireless cabin including accommodation, supply the electric current and maintain the masts. The Marine Company was to be enabled to carry on telegraphic business with all passengers and persons on board, and was to receive and transmit free of charge (up to a limit, beyond which a charge at half the tariff rate applied) messages having relation to the navigation of or any business connected with the ships passing between the Masters of the said ships and the shipowners or

next page
The Marconi room on the *Olympic*.

other Masters (messages in this category were referred to as 'Masters' service messages'). The operators were to be signed on the ship's Articles, so they would be under the Master's orders although employed by the Marine Company.[2]

An article in the Marconi Company's magazine *The Marconigraph* shortly after the disaster provides a summary description of the equipment installed on the *Titanic*:

> The wireless equipment of the *Titanic* was the most powerful possessed by any vessel of the mercantile marine, and only equalled by that of the *Olympic*. Its generating plant consisted of a 5-kw. motor-generator set, yielding current at 300 volts 60 cycles. The motor of the set was fed at 110 volts d.c. from the ship's lighting circuit, normally supplied from steam-driven sets; while, in addition, an independent oil-engine set was installed on the top deck, and a battery of accumulators was also provided as a stand-by. The alternator of the motor-generator set was connected to the primary of an air-core transformer, and the condenser consisted of oil-immersed glass plates. To eliminate as far as possible the spark-gap and its consequent resistance, which, as is well known, is the principal cause of the damping of the waves on the transmitting circuit, the ordinary Marconi rotary disc discharger was used. This is driven off the shaft of the motor-generator. The guaranteed working range of the equipment was 250 miles under any atmospheric conditions, but actually communication could be kept up to about 400 miles, while at night the range was often increased to about 2,000 miles. The aerial was supported by two masts, 200ft. high, stepped 600ft. apart, and had a mean height of 170ft. It was of the twin T type, and was used for the double purpose of transmitting and receiving. The earth connection was made by insulated cable to convenient points on the hull of the vessel.

The receiver was the Marconi standard magnetic travelling band detector used in conjunction with their multiple tuner, providing for the reception of all waves between 100 and 2,500 metres. The multiple tuner was calibrated to permit of the instrument being set to any prearranged wavelength, and further to be provided with a change switch to permit of instantaneous change of the circuit from a highly syntonised tuned condition to an untuned condition (for standby) especially devised for picking up incoming signals of widely different wavelengths. By reason of its robust nature the magnetic detector could be employed permanently connected to the transmitting aerial, thus dispensing with all mechanical changeover switching arrangements.[3]

The equipment described was the most advanced and powerful of any to be found, as befitted a ship of the status of the *Titanic*, and matched only by that on her sister ship, the *Olympic* launched a year earlier. There is only one surviving photograph of the wireless cabin on board the *Titanic*: that taken by a Jesuit priest, Father Browne, who captured several images of the ship before disembarking at Queenstown in Ireland. This photograph, which was double-exposed, shows a rear view of Harold Bride – one of the *Titanic*'s radio operators – working at his equipment.

Images of the aerial are also elusive, as it is not apparent in several of the surviving images of the ship. One in *The Marconigraph* accompanying the above description shows its four parallel wires remarkably clearly, suspended between the two masts on the deck.

The *Titanic* leaving Southampton. The aerial for the wireless equipment is clearly visible. (Oxford, Museum of the History of Science, *The Marconigraph*, vol. II, p. 40.)

Harold Bride at work in the *Titanic*'s Marconi room

THE RADIO OPERATORS

An article in the first volume of *The Marconigraph* describes the requirements, training and life of a Marconi radio operator in late 1911, such as would be experienced by those on ships involved in the disaster and its aftermath. In part this was a plug for new recruits, and sought to present an attractive view of an operator's career while being careful to seek out individuals

BIAGHI
GENOVA
MARCONI'S
WIRELESS TELEGRAPH C
INSTALLATION FITTED ON
CUNARD Co's PASSENGER STEAME
MESSAGES
RECEIVED HERE
FOR TRANSMISSION TO SHIPS
IN THE ATLANTIC
NOTICE TO PASSENGERS
Marconi International Marine
Communication Company Ltd
TELEGRAPHIC COMMUNICATION BETWEEN SHIP AND SHORE
18 FINCH LANE LONDON E.C.

suited to the work. In terms of qualifications, men between 21 and 25 years old were eligible, preferably with previous experience of inland or cable telegraphy. Individuals had to be able to send and receive at the rate of not less than twenty-five words per minute on a Morse key and sounder. Knowledge of magnetism and electricity was necessary. Approved applicants were sent to the Company's Liverpool training school for a course of instruction in wireless telegraphy, sufficient to prepare them for the examination held by the Postmaster-General, who issued certificates of competency to successful candidates. The Liverpool school was at Crosby, Seaforth, having recently moved to new premises there after being established in 1903 at Seaforth Sands.

The training course incorporated several elements of study: elementary electricity and magnetism; fundamental principles of wireless telegraphy; transmitting by practice buzzer sets and receiving by telephones as used in wireless; the various pieces and types of apparatus used, and diagrams of electrical connections thereof; the connecting up of various parts comprising complete sets, how to trace and remove faults and repair breakdowns; rules and regulations laid down by the Radiotelegraph Convention for the commercial working of wireless telegraphy; clerical work in connection with telegraphic accounts and returns; and general routine and discipline on board ship.

The wireless operator on board ship signed on to the ship's articles as a member of the crew, generally with the honorary rank of a junior officer, and was subject to the disciplinary

opposite
Replica of a Marconi wireless cabin on a Cunard steamer, at a trade exhibition stand, c. 1910. (Oxford, Bodleian Library, MS. Photogr. c. 253, item 340/X.)

The Marconi International Marine Communication Company's *General Orders*, 1913. (Oxford, Bodleian Library, MS. Marconi 244, cover.)

regulations of the ship. On ships with two operators a continuous watch was maintained; if there was just one operator then the hours of duty were long and irregular. In terms of remuneration, an operator commenced as a junior. The rate of pay was said to be slightly higher than that of telegraphists in the home government and cable services, and much higher than railway telegraphists; an added attraction was that it

Reg. 60.

CHAPTER IX.

SIGNALLING—METHOD OF.

(60.) Below is given the International Morse Code Alphabet and Numerals.

Alphabet.

A	· —	H	· · · ·	Q	— — · —
Ä	· — · —	I	· ·	R	· — ·
Á or Å	· — — · —	J	· — — —	S	· · ·
B	— · · ·	K	— · —	T	—
C	— · — ·	L	· — · ·	U	· · —
Ch	— — — —	M	— —	Ü	· · — —
D	— · ·	N	— ·	V	· · · —
E	·	Ñ	— — · — —	W	· — —
É	· · — · ·	O	— — —	X	— · · —
F	· · — ·	Ö	— — — ·	Y	— · — —
G	— — ·	P	· — — ·	Z	— — · ·

Numerals.

1	· — — — —	5	· · · · ·	9	— — — — ·
2	· · — — —	6	— · · · ·	0	— — — — —
3	· · · — —	7	— — · · ·		
4	· · · · —	8	— — — · ·		

The following signals may also be employed to express Figures, but only in official repetitions and in the preamble, and in the text of telegrams written entirely in figures :—

1	· —	5	·	9	— ·
2	· · —	6	— · · · ·	0	—
3	· · · —	7	— · · ·		
4	· · · · —	8	— · ·		

— 42 —

Page from the *General Orders* detailing the Morse Code. (Oxford, Bodleian Library, MS. Marconi 244, p. 42.)

increased more rapidly. The life was varied and interesting, with 'all sorts of special appointments on board large steam yachts, foreign warships and even airships; also special work on land in foreign countries'.

On many ships the operator received nightly from high-power shore stations the condensed news of the world for publication on board his ship. This would commonly be inserted into mostly preprinted 'ocean newspapers', with space to insert the news items received. From other ships he obtained first-hand messages containing information as to meteorological conditions, ice reports, and the presence of fog and of derelicts.

The article ended with some words of both encouragement and warning:

> To the man who enters the service without an exaggerated notion of his own importance, or whose head is not likely to become swollen by the responsible nature of his work and the donning of uniform, there are many opportunities for advancement.... Let him not for one moment suppose that in the operating side of wireless telegraphy he has discovered a short and easy way to an 'El Dorado'; but without painting an alluring picture ... the attractions and the remuneration are such as make a favourable bid for the entry of capable and zealous young men. There is plenty of room in the service for the right sort of worker, but for the shirker, none – absolutely none.[4]

The *Titanic* had two radio officers on board, John (Jack) Phillips and Harold Bride. Jack Phillips, aged 25, came from Farncombe near Godalming in Surrey. He started his career as a telegraph learner in the Godalming Post Office, and joined the Marconi Company as a learner at the Liverpool School in March 1906. Phillips was appointed to the staff on 6 June 1906 and sailed on several ships, including the SS *Teutonic*, SS *Pretorian*, SS *Buccaneer* and RMS *Oceanic*. He also served for three years at the high-power transatlantic wireless station at Clifden, Ireland (where he worked transmitting and receiving

Jack Phillips, senior operator on the *Titanic*, who perished in the disaster. (Oxford, Bodleian Library, MS. Marconi 256, fol. 148.)

messages from the Marconi station at Glace Bay, Nova Scotia) between May 1908 and July 1911. He rose to the position of Senior Operator, in which position he served on the *Titanic*. The radio operators' staff books (apparently compiled at a later date) do not include entries for Phillips or those of his key colleagues involved in the disaster, nor do their record cards survive, but there are some details of their service in a staff salary book in the Marconi Archives. Phillips's short entry concludes poignantly, with a note of his drowning on 16 April 1912 on the *Titanic*.[5]

44

Name	Date joined + Dates of increases	Increase	Salary from 1 July '11			Date & reason of leaving	Remarks
Standbridge. S. Sidney	6 June '06		~~1~~	~~5~~	·		
			1	10	0		Inc: from 30/6/12.
							* 3/11/08
	1st July 1913	15/-	2	5	0		Appointed, Travelling Inspector 5/10/16
	5th Oct 1913	5/-	2	10	0		
	3rd Jan 15	5/-	2	15	0		
	2nd April 16	5/-	3	0	0		
Phillips J G	6 June '06		1	15	·	16th Apr 1912	R M S Titanic Drowned

Entry for Jack Phillips from staff salary book, 1912. (Oxford, Bodleian Library, MS. Marconi 2040, p. 44.)

Harold Bride, the junior operator on the *Titanic*, was born at Nunhead, near Greenwich, in 1890. Trained at the British School of Wireless Telegraphy at Clapham, London, he was appointed by the Marine Company on 3 July 1911 as a learner and joined the staff as a qualified operator on 6 August 1911. He served on the SS *Haverford*, RMS *Lusitania* and RMS *Anselm* before being assigned to the *Titanic*. He survived the sinking and continued as a radio operator for a year or so. His salary record indicates that in October 1913 he took two years' leave of absence without pay, then returned in May 1915 on 'special duty at Leafield station and room 106'.[6] He finally resigned the service in May 1917.[7] He died in 1956.

above
Harold Bride, junior operator on the *Titanic*. (Oxford, Museum of the History of Science, *The Marconigraph*, vol. II, p. 40.)

Harold Cottam, the sole radio operator on the *Carpathia*, was the other Marconi man to play a major part in the disaster. Born in 1891 in Southwell, Nottinghamshire, he was appointed as an operator on 9 May 1910. Like Bride he continued in service after the sinking of the *Titanic*, and in 1915–18 undertook war service in the form of special duty under Lieutenant Round, who during the First World War was involved in the provision of Admiralty wireless stations and headed work on direction finding by wireless. Cottam resigned from the company in December 1923, and died in 1984.[8]

below
Harold Cottam, operator on the *Carpathia*. (Oxford, Museum of the History of Science, *The Marconigraph*, vol. II, p. 103.)

Other wireless operators who played a role included Cyril Evans on the SS *Californian*, Gilbert Balfour on the RMS *Baltic*, C.J. Moore and A. Boyd on the *Olympic*, Stanley Adames on the SS *Mesaba*, and J. Durrant on the SS *Mount Temple*.

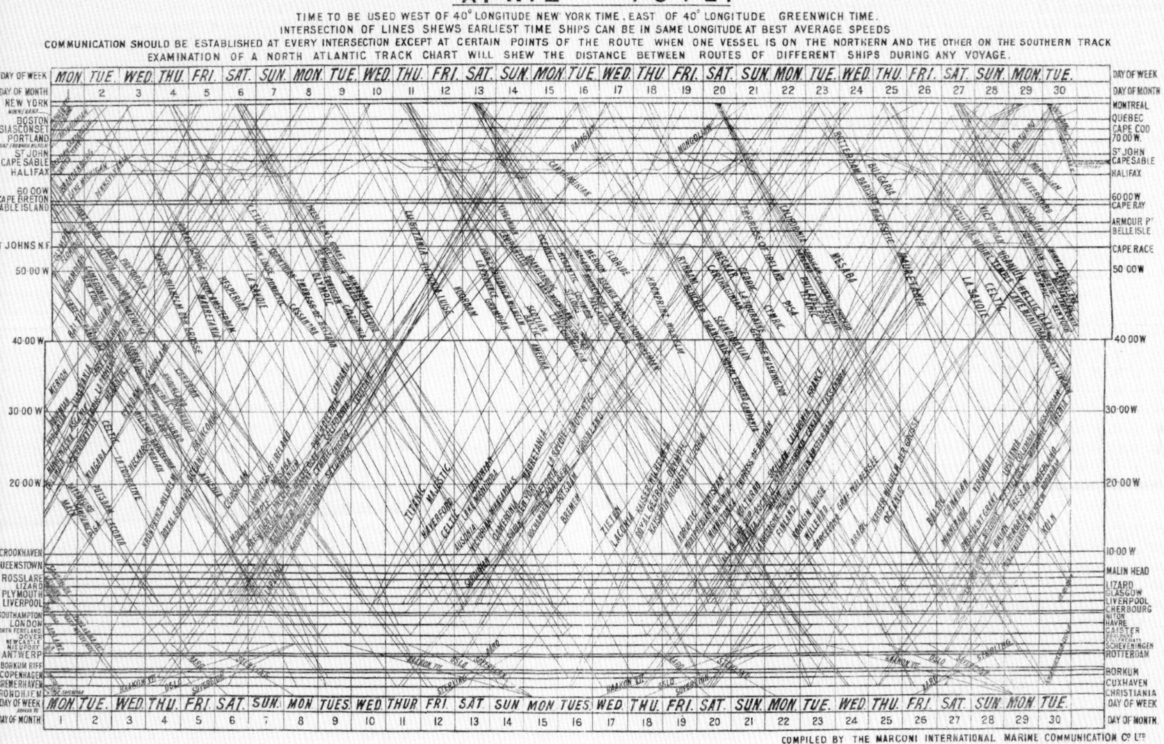

Track and Communication Chart for the North Atlantic, April 1912. (Oxford, Bodleian Library, MS. Marconi 258, fol. 202.)

At the time of the *Titanic*'s maiden voyage, various operational routines had developed for maritime wireless telegraphy. Wireless operators on both ship and shore stations kept records of the messages received and sent. Full transcripts of messages handled were made on forms provided for the purpose, or written out by passengers on similar forms for transmission. The station's procès-verbal summarised activities in chronological order, only giving full details of the content of a message if the operator considered it worthy of note. Standard codes were used to abbreviate the names of stations called: for example, the code for the *Titanic* was MGY, and Marconi stations were all prefixed M. Monthly Track and Communication Charts were available for the principal shipping routes to estimate when other vessels equipped with wireless would be within calling range. It was the practice to exchange batches of messages to or from the ship with the principal shore stations when they came within range, often at night when the range was much greater, and those destined for recipients on land would be forwarded by the coastal wireless stations, either by wireless or by cable. Messages ordered by the captain to be sent to his counterpart on another vessel were known as Masters' Service Messages and had to be taken straight to the bridge upon receipt.

THE TRANSATLANTIC VOYAGE: ICE WARNINGS AND PASSENGERS' MESSAGES BEFORE THE EMERGENCY

The *Titanic* set out from Southampton on its maiden voyage on Wednesday 10 April 1912. After calling at Cherbourg and Queenstown to pick up further passengers (and allow a few to disembark, including Father Browne, mentioned above), the ship began steaming across the Atlantic in earnest on the afternoon of 11 April.

Guglielmo Marconi and his wife Beatrice had been invited to join the maiden voyage as guests of the White Star Line. They accepted, but then had to withdraw because Marconi found that for business reasons he had to reach the United States earlier than scheduled, and instead travelled on the *Lusitania*. Beatrice also changed her plans, deciding to remain behind altogether due to the illness of their son.[9]

A book of messages in the Marconi Archives contains several communications relating to the voyage prior to the point at which the iceberg was struck. Somewhat confusingly entitled 'Messages relating to ice in the vicinity of the *Titanic*', its contents cover other topics besides. There are greetings sent to the *Titanic*'s Captain Smith congratulating him on his command of this state-of-the-art vessel, and later messages covering the emergency itself and the efforts of other ships to aid the sinking *Titanic*. However, many are the warnings typically exchanged between ships traversing the North Atlantic, giving data on the location of icebergs and sheet ice so that these could be avoided:

> Congratulations on new command had moderate westerly winds fair weather no fog much ice reported in Lat 42.24 to 42.45 and Long 49.50 to 50.20 Compts.[10]

> Westbound steamers report bergs growlers and field ice in 42 N from 49 to 51 West April 12.[11]

The messages in this book (which survives in duplicate) are stamped 'copy', suggesting that they were transcribed at the time from the original records, for use in the British Inquiry. Their original numbering in blue pencil corresponds with the way in which certain messages were described as being marked up for presentation to the Inquiry.

There are examples of Captain Smith's responses acknowledging these friendly warnings, but nothing to indicate direct actions as a result. The ice warning messages were of considerable interest to the Inquiry and are discussed further below.

A little more detail about the process of sending ice messages may be gleaned from other documents, such as the statement of Stanley Adames, operator on the *Mesaba*, annexed to the *Mesaba*'s procès-verbal. He describes how he sent an ice report for all eastbound ships and also to the *Titanic*. Adames states that the *Titanic* could not hear until the ships got closer, the message being successfully transmitted at 7.50 p.m. New York time, and according to Adames acknowledged by the *Titanic* in the usual way – a significant statement as it is not clear from the *Mesaba* PV that it was the *Titanic* replying.[12]

Some messages to or from passengers on the *Titanic* are to be found in other series of messages in the collection. Notable examples are those between Dorothy Gibson, a Hollywood film actress and model on board the *Titanic*, and her lover Jules Brulatour, a film entrepreneur, in New York. Jules sent messages to Dorothy on her voyage, including the following sent via Cape Cod station on 12 April:

> Will do everything make you completely happy Love you madly. Julie.[13]

Further messages between Jules and Dorothy are given in the transcripts of personal messages. More will be heard of this couple a little later.

THE ICEBERG IS HIT

At 11.40 p.m. (ship's time) on 14 April, the *Titanic* struck an iceberg and a hole estimated at some 300 feet in length was torn out of her side beneath the water line. The bulkheads failed to limit the flooding, and the vessel began to sink. The wireless operator on duty, Jack Phillips, was instructed to send out distress signals, which he began to do at 12.05 a.m.

The *Mount Temple* was probably the first to hear the *Titanic*'s emergency call, an entry in its procès-verbal noting the CQD and request for assistance. The *Mount Temple* replied, but could not be heard by the *Titanic*. News of the emergency spread rapidly to other ships in the vicinity, and several responded, including the *Frankfurt* and the *Carpathia*.

Ensuing events are recorded by way of the procès-verbaux of the ships involved, coupled with messages in the ice warnings book. Typical distress messages were picked up by the SS *La Provence* and the RMS *Celtic*:

> Position 41.46 N 50.14 W require assistance struck iceberg[14]
>
> CQD require assistance position 41.46 N 50.14 W struck iceberg Titanic[15]

Both CQD and SOS were sent out as distress signals. CQD was the emergency call sign of Marconi operators and would have been widely recognised; SOS had recently been introduced as the international distress call and possibly was less familiar, though should have been understood by all operators.

Attempts were made to rouse the *Californian*, thought to be the vessel nearest the *Titanic*. However, the *Californian* was out of wireless communication since its sole radio operator had retired for the night. There is considerable controversy about the *Californian* because of her proximity to the *Titanic*, in fact being so close that the lights of the *Titanic* could be seen but were not recognised as those of a liner. Also, the watch reported seeing rockets being fired, but these were not recognised as distress signals. Unsuccessful attempts were made by both vessels to communicate by Morse lamp, but the wireless operator was not roused to seek contact by that means.[16]

The *Californian* was eventually made aware of the sinking, the operator Cyril Evans being awoken at 3.30 a.m. He then communicated with the SS *Virginian*:

> Please give MSG on account MGY [Titanic] so as capt can go off track down to MGY.[17]

The *Virginian* replied:

> Titanic struck bergs wants assistance urgently ship sinking passengers in boats his position Lat 41.46 Long. 50.14[18]

The superior power of the *Titanic*'s sister ship, the *Olympic*, is clear from her ability to communicate from some 500 miles away. The *Olympic* made full steam for the stricken vessel, her captain sending the message:

> Am lighting up all possible boilers as fast as can.[19]

However, it became clear that other ships could reach the site of the sinking much sooner. In the event it was the *Carpathia* alone that took on this role, but the written record shows the confusion surrounding events, as some messages from survivors on board the *Carpathia* refer to the possibility of survivors being picked up by other ships, which was not the case.

The procès-verbal of the *Virginian* provides a compelling and poignant record of the last known messages to be sent from the *Titanic*. The *Virginian*'s operator heard CQD and SOS messages, but communication was breaking up. Two V's were then heard, in a signal like the *Titanic*'s, as if the operator was adjusting the spark. Finally power was abruptly cut off – this is the point at which the *Titanic*'s wireless room lost its power:

> 12.27 am MGY calls CQ unable to make out his signal Ended very abruptly as if power suddenly switched off His spark rather blurred or ragged Called MGY & suggested he shd try emergency set but heard no response.[20]

The official time at which the *Titanic* foundered was 2.20 a.m. (*Titanic* time). The *Carpathia* was the first vessel to reach the scene of the disaster, arriving at daybreak. By that time the *Titanic* had sunk and the *Carpathia* picked up survivors from the lifeboats, being careful to account for all the boats.

AFTER THE SINKING: THE *CARPATHIA* HEADS TO NEW YORK

After the *Carpathia* had picked up all survivors, it began steaming towards New York. Other ships which had altered course towards the coordinates of the sinking were stood down, as revealed by a message from the *Carpathia* to the *Virginian*, recorded in her procès-verbal:

> At 8am Carpathia said:–
>
> 'Tell your Skipper we are leaving here with all on board – about 800 passengers Everything OK Please return to your northern course.'[21]

One ship that did not return to its course was the *Californian*, which headed to the scene to search the neighbourhood after the *Carpathia* had left.[22]

The *Olympic* sought information from the *Carpathia* as to its best course of action:

> Shall I meet you and where.[23]

The *Carpathia* cautioned against a rendezvous, apparently concerned that the *Titanic* survivors might be further distressed by the sight of the similar form of the sister ship. The *Olympic* concurred, but remained anxious to continue searching for survivors if possible, asking whether there was the slightest hope in searching the *Titanic*'s position, and whether news of the disaster had been communicated to New York or Liverpool. The *Carpathia*'s next response summarised the position:

Fear absolutely no hope searching Titanics position left Leyland SS Californian searching round all boats accounted for about 675 souls saved latter nearly all women and children Titanic foundered about 2.20am 5.47 GMT in 41–16 N 50–14 W not certain of having got through please forward to White Star also to Cunard Liverpool and New Yk and that I am returning New Yk consider this most advisable for many considerations[24]

During the period of the *Carpathia*'s voyage to New York, there was intense wireless traffic relating to the disaster. The *Carpathia* itself focused on sending messages from survivors to their relatives, and communicating lists of those survivors. Harold Cottam was assisted in this task by the surviving operator from the *Titanic*, Harold Bride, this despite Bride having spent some time in the water before managing to get onto a collapsible lifeboat. The survivors' personal messages are well represented in the numerous service forms in the series of messages handled by the shore stations while the *Carpathia* was heading towards New York. Some of these examples bring home the full force of the tragedy:

To Lewis Peake, Paris: 'Safe board Carpathia bound to New York. Mother.'

To Mr. H.G. Berry, New York: 'Safe on Carpathia Holrey lost meet me. Neate.'

To Rheinis, New York: 'Meet me dock with two hundred dollars under wear coat am well but feet slightly frozen answer. George.'

To Lucilation, London: 'Sir Cosmo and Lady Duff Gordon safe. Carpathia.'

To Margaret Morris Brown, London: 'Jack Margaret are safe no news Johnny. Marian Thayer.'

To Percy Strauss, New York: 'Every boat watched father mother not on Carpathia hope still. Badenoch.'

To Norris Williams, Philadelphia: 'Father not seen no hope arrive Carpathia Wednesday New York. Richard.'

To Simon Madigan, Askeaton, Ireland: 'The ship sunk B.M. and I are safe.'[25]

The difficulties individuals experienced in obtaining news are exemplified by Jules Brulatour's attempts to contact Dorothy Gibson, trying on numerous occasions to send a message to find out what had happened to his lover. These messages were stamped 'not transmitted' because of the inability of the Camperdown station to get in contact with the *Carpathia*. But Dorothy seems to have been alerted eventually, or at any rate contacted Jules independently, for the following message was eventually despatched:

Safe picked up by Carpathia dont worry. Dorothy.[26]

Gibson reached New York and her lover safely, and the two were eventually married. Before that, Dorothy starred in the first film about the sinking of the *Titanic*, produced by Brulatour later in 1912. Brulatour later became co-founder and first president of Universal Pictures.

The procès-verbal of the SS *Minnewaska* and a few messages annexed to it illuminate efforts to obtain lists of survivors at this time. The New York office of White Star Line contacted the *Minnewaska*, which was within its range:

> We are particularly anxious ascertain from Carpathia complete list of names of survivors Titanic aboard Carpathia Do your utmost get in touch with her and if so spare no trouble to get us all names and wire them unless you find Carpathia has forwarded them to us answer Franklin.[27]

The *Minnewaska* eventually succeeded in conveying this to the *Carpathia*, and elicited a response from Captain Rostron:

> Only too anxious to get all names to shore doing all possible Mr Ismay is on board bergs and pack ice down to 41.16 N and as far east as 49.30 W.[28]

The *Minnewaska*'s procès-verbal brings home the difficulties encountered in transmitting the list:

> 1.30 pm MPA will not send traffic to MSD & insists on sending list of names to me saying they are for our captain. I give him g & he starts. There are several repetitions owing to jamming & somewhat to poor transmission. Operator on MPA is evidently tired out. Finally he understands that I am unable to get MSD or rather that MSD cant read me & stops sending.
>
> 5.10 pm On return hear MPA wkg MSD Evidently there has been a change of operator. Expect one off Titanic. Standing bi with msg for MWL.[29]

The frustration of Cottam himself is brought home by a letter he felt compelled to write to G.E. Turnbull, deputy manager of the Marine Company in London, in which he complained vehemently about the SS *Birma*, which

> made himself a general nuisance the whole time I was within his range as is proved by the S/S Baltic's PV asking silly unnecessary questions. … Captain Rostron told me to ignore all ships other than those which I should benefit by working.[30]

By the time the *Carpathia* was close to New York, thoughts were turning to what was to happen on shore and a message was sent by Cottam to find out what he should do:

> Please wire instructions for second operator ex Titanic and myself Titanic man unable to walk regret unable to cope with enormous rush of work. Opr.[31]

Bruce Ismay, chairman of the White Star Line and a survivor of the sinking, had already been in touch with his company in New York. He had sent a formal notice of the sinking:

> Deeply regret inform you titanic sank this morning after collision iceberg resulting serious loss life further particulars later. Bruce Ismay.[32]

Ismay then had a series of exchanges with his deputy, P.A.S. Franklin, one conveying a message of relief at his survival and sympathy over the disaster from Ismay's wife, Florence. Ismay sought to make arrangements for dealing with the *Titanic*'s lifeboats on arrival at quarantine in New York, and for the rapid return of the *Titanic*'s crew to Britain by holding the RMS *Cedric* in New York for the purpose. These messages are detailed in the selection of official messages transcribed below. Ismay signed most of the messages with his name spelled backwards.

Other passengers took the initiative to make further practical arrangements, Leila Meyer and Julia Siegel ordering quantities of clothing from their families' stores in New York (Saks & Co. and Siegel-Cooper & Co.) for the destitute among the survivors.[33]

GLOBAL COMMUNICATION OF THE DISASTER

In the hours and days following the moment of striking the iceberg, information filtered and then flooded beyond the immediate vicinity of the emergency. The bare facts of the news spread remarkably quickly, reflecting the high power of the *Titanic*'s wireless equipment and its capacity to reach the coastal stations of North America, and even further afield. The spread of news also led to frustration and confusion: the head office of the Marine Company in London complained to the Camperdown shore station that Reuter's was getting news an hour earlier than their source, and ironically urged its colleagues to use cable to provide a better service to them. Camperdown's reply commented that it had no news to give. Later, it did enlarge upon this:

> Absolutely swamped Carpathia traffic Franconia making poor progress relaying Carpathia operator over two hundred messages remain to clear ... doing utmost obtain officially information you seek. 705 survivors Carpathia rest missing Carpathia expects arrive port Thursday evening.[34]

There was also concern from Southampton, the base of the White Star Line, seeking news of its crew:

> Tension on relatives awaiting news very acute harrowing scenes amongst crowd awaiting information.[35]

This was passed on to Camperdown and Siasconset shore stations in an attempt to get information. Siasconset replied:

> Operator Carpathia advises impossible compile list crew survivors wireless clogged with traffic.[36]

Meanwhile, the Newfoundland government was greatly annoyed at its inability to obtain news of the disaster from Cape Race station, and the governor contacted the Marine Company in London in an attempt to resolve the situation. The

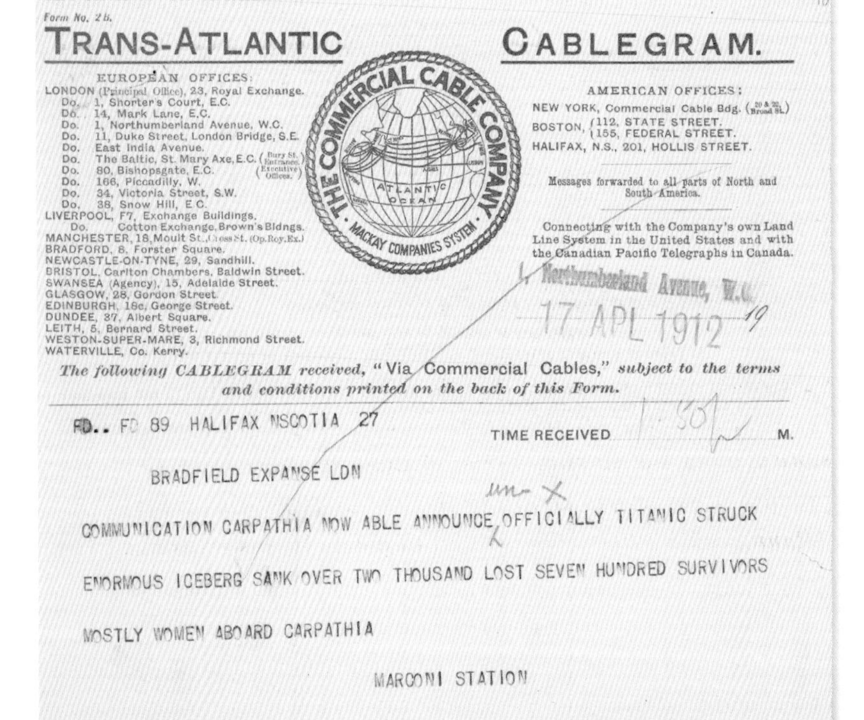

10

Form No. 26.

TRANS-ATLANTIC CABLEGRAM.

THE COMMERCIAL CABLE COMPANY · MACKAY COMPANIES SYSTEM · ATLANTIC OCEAN

EUROPEAN OFFICES:
LONDON (Principal Office), 23, Royal Exchange.
Do. 1, Shorter's Court, E.C.
Do. 14, Mark Lane, E.C.
Do. 1, Northumberland Avenue, W.C.
Do. 11, Duke Street, London Bridge, S.E.
Do. East India Avenue.
Do. The Baltic, St. Mary Axe, E.C. (Bury St. Entrance.)
Do. 80, Bishopsgate, E.C. (Executive Offices.)
Do. 166, Piccadilly, W.
Do. 34, Victoria Street, S.W.
Do. 38, Snow Hill, E.C.
LIVERPOOL, F7, Exchange Buildings.
Do. Cotton Exchange, Brown's Bldngs.
MANCHESTER, 18, Moult St., (Cross St.) (Op. Roy. Ex.)
BRADFORD, 8, Forster Square.
NEWCASTLE-ON-TYNE, 29, Sandhill.
BRISTOL, Carlton Chambers, Baldwin Street.
SWANSEA (Agency), 15, Adelaide Street.
GLASGOW, 28, Gordon Street.
EDINBURGH, 18c, George Street.
DUNDEE, 37, Albert Square.
LEITH, 5, Bernard Street.
WESTON-SUPER-MARE, 3, Richmond Street.
WATERVILLE, Co. Kerry.

AMERICAN OFFICES:
NEW YORK, Commercial Cable Bdg. (20 & 22 Broad St.)
BOSTON, 112, STATE STREET. 155, FEDERAL STREET.
HALIFAX, N.S., 201, HOLLIS STREET.

Messages forwarded to all parts of North and South America.

Connecting with the Company's own Land Line System in the United States and with the Canadian Pacific Telegraphs in Canada.

1, Northumberland Avenue, W.C.
17 APL 1912

The following CABLEGRAM received, "Via Commercial Cables," subject to the terms and conditions printed on the back of this Form.

FD.. FD 89 HALIFAX NSCOTIA 27 TIME RECEIVED M.

BRADFIELD EXPANSE LDN

COMMUNICATION CARPATHIA NOW ABLE ANNOUNCE OFFICIALLY TITANIC STRUCK ENORMOUS ICEBERG SANK OVER TWO THOUSAND LOST SEVEN HUNDRED SURVIVORS MOSTLY WOMEN ABOARD CARPATHIA

MARCONI STATION

Cable telegram from the Halifax wireless station to the Marconi Company in London on 17 April, unofficially confirming the sinking of the *Titanic*. (Oxford, Bodleian Library, MS. Marconi 256, fol. 10.)

Marine Company sought to appease and asked its Canadian counterpart to instruct Cape Race to do what it could, while explaining to the governor that Cape Race might not be in receipt of definite information, and that what information was available was being published in the press.[37]

A more particular source of confusion during the aftermath was the activities of the SS *Asian* on 14 April, as reported in her procès-verbal:

> 7 am Stopped to tow German tank str (39.40 N 50.20 W) Deutschland to Halifax.[38]

Communication of this event appeared to lead to rumours that the *Titanic* had not sunk but was being towed to Halifax.

> Thanks your message we have nothing from Titanic but rumored that she proceeding slowly halifax but we cannot confirm this we expect Virginian alongside Titanic try communicate her.[39]

The confusion was such that the Marine Company printed extracts from the *Asian*'s procès-verbal and related messages for the Inquiry, to clarify what had taken place. The printed version is annexed to the *Asian*'s procès-verbal.[40]

The press interest in the sinking was, of course, considerable. The records include a separate series of messages handled by Camperdown from journalists seeking news and personal stories about the sinking. Some of these related to specific prominent passengers known or thought to have been on the *Titanic*, while others endeavoured to obtain exclusives from crew about their experiences:

> Five hundred dollars for column and half more in proportion please rush to get here before midnight would like good story also after you land.[41]

> Please wireless all you can of the Titanic will pay at rate one hundred fifty dollars a column.[42]

> Kindly wireless fullest account Titanic disaster Greatly relieve world wide anxiety desire especially know fate Astor Stead Strang Millett Cutts and other famous persons.[43]

> Please send story shipwreck wireless or otherwise shipwreck unlimited quantity.[44]

> Please rush definite information regarding loss of life names of survivors allay anxiety of thousands here send wireless collect any length.[45]

Once again, considerable frustration reigned as many of these messages could not be transmitted by Camperdown because of the heavy volume of higher-priority wireless traffic. Despite this, coverage was extensive and the large series of press cutting albums in the Marconi Archives includes three substantial volumes devoted to the calamity.[46]

AFTERMATH: GATHERING THE EVIDENCE

Within a few days of the disaster, the Marine Company in London began systematically gathering the written records of its wireless stations in order to present as full a picture as possible to the British Inquiry into the sinking. On 18 April, it wrote to the Marconi Wireless Telegraph Company of America urgently requesting logs for South Wellfleet Siasconset and Sagaponack stations also message copies from 14th to 17th inclusive also same from Carpathia and all other incoming steamers.[47]

A similar message was sent to the Marconi Wireless Telegraph Company of Canada, asking for records of Cape Race, Camperdown and Cape Sable stations, together with those for the SS *Parisian* and other incoming steamers. Associated companies in Belgium, France and Germany were also contacted. It sought in particular the records of any messages transmitted between vessels pertaining to ice. The companies were quick to oblige as rapidly as was within their power, but there were inevitable delays as the documentation had to be despatched by ships returning to Britain. London emphasised the need for the original records, but the American company for example sent copies at first, resulting in a follow-up telegram urgently requesting the originals. Another complication was that the captain of the *Carpathia* refused to allow the station's records to be taken off the ship, resulting in a request from the Marine Company to the Cunard Company to instruct their captain to comply, and allow the evidence to be despatched upon reaching Gibraltar on its return voyage. The *Carpathia*'s procès-verbal was eventually acquired and is among the other PVs in the Marconi Archives. It is unusual, however, in ending abruptly at a point shortly before distress signals started to be received: the operator (even when later assisted by his colleague from the *Titanic*) was too busy sending and receiving messages to keep a log from that point onwards.

In the Marconi Archives there is a significant series of detailed correspondence between the Marine Company and Sir R. Ellis Cunliffe, Solicitor to the Board of Trade, and his officials, concerning the nature of the evidence required from the Company for the Inquiry. These discussions continued while the Inquiry was in progress – in part because certain evidence was not yet to hand, being still in transit back across the Atlantic. The Company's staff were likewise travelling back to give evidence in person. The correspondence with Cunliffe mostly went out under the general signature of the Marine Company: it was probably the deputy manager, G.E. Turnbull, who undertook the correspondence. Turnbull appears to have been the person drawing together all the material on behalf of the Company, and himself presented evidence to the UK Inquiry.

A letter of 1 May (later printed for the benefit of the Inquiry) gives an early report on the Company's efforts to collect the available evidence. At that point, records of the US stations had been received, but not those from Canadian or European Marconi companies. The records of the *Carpathia* were also still awaited from Gibraltar. The letter sets out the nine ship stations then known to have picked up the *Titanic*'s distress signals. The printed letter annexes extracts from the reports of the *Olympic*, *Baltic*, *Mount Temple*, *Caronia*, *Virginian*, *Asian* and *Parisian* from the time of the first distress signal until shortly after the *Titanic* foundered. At this point it seemed there was a greater emphasis on the events following the distress call,

180

The Marconi International Marine Communication Co. Ltd.,

MARCONI HOUSE, STRAND, LONDON, W.C.

Substitute for the *proces verbaux* of the "TITANIC," from the 12th April until her last signals were received; also for the "CARPATHIA," from the 14th to 18th April, 1912.

The "TITANIC" Telegraph Records are lost, and the "CARPATHIA'S" *proces verbal* was not continued after the "TITANIC'S" distress call was first received on the "CARPATHIA."

In order to take the place as far as possible of the missing *proces verbaux* this Document has been compiled from the Records of other Wireless Stations which either communicated direct with these two Vessels or overheard communications passing with them or concerning them.

The portion of this Document relative to the communications effected by the "TITANIC" commences on the Twelfth of April, at 12.21 p.m. Greenwich Mean Time (G.M.T.), and concludes at 12.27 a.m. New York Time (N.Y.T.) on the Fifteenth of April, with the last Wireless Signal heard from her as reported by the "VIRGINIAN." The day-divisions in this document are made by New York Time.

The References to the "CARPATHIA" commence at the time this Ship first communicated with the "TITANIC" on the Fourteenth of April, at 5.10 p.m. New York Time, and continue up to the time the "CARPATHIA" arrived at New York with the Survivors on the Eighteenth of April at 9.45 p.m.

COMPILED, LONDON, 8th JUNE, 1912.

Substitute for the procès-verbaux of the *Titanic* and *Carpathia*, compiled for the British Inquiry. (Oxford, Bodleian Library, MS. Marconi 258, fol. 180.)

rather than upon the evidence for ice warning messages being sent or received, which became a particular focus of interest of the Court of Inquiry.[48]

A further letter of 7 May, also prepared in printed format, provides an update. The records of the *Carpathia* were now to hand, and a copy of the procès-verbal covering the period 14–18 April is annexed to the printed version. This is somewhat puzzling as the original procès-verbal ceased just before receipt of the distress signals: this may be an early version of the substitute procès-verbal compiled from other stations' records, although this is not stated and the existence of an original is implied. Also annexed are copies of four messages relating to ice taken from records submitted by the wireless companies, exemplifying the increasing interest in this aspect of the disaster.[49]

A draft of Turnbull's intended statement (undated but presumably before 22 May, when he began his evidence in person) provides insights into the conduct of business by Marconi operators on board ship:

> The system whereby messages are sent and received by an Operator and recorded by him is as follows:–
>
> The wireless Operator is sitting during his hours of duty with the receiver to his ears. As soon as he receives a call directed to his own ship he replies, giving the vessel in communication with him a note to proceed. He then on receipt of a message simultaneously with the receipt writes it down as he receives it, generally on an official form over carbon paper though in some cases on a separate slip of paper. If he writes it on a separate

slip of paper it is transcribed on to an official form at the first opportunity, over carbon paper, and later the carbon copy is delivered to the addresses on board ship and the original is retained and subsequently sent, at the end of the voyage, to the head Office of the Company in whose employ he is.

As regards messages sent by the wireless Operator, the Captain or any other Officer or any passenger, who is sending the message, himself writes it out on the official form as is done in land telegraph offices, and this is the form which is kept by the wireless Operator and is sent, at the end of the voyage, to the Wireless Company which controls the Stations on his vessel. The wireless Operator also copies down in a similar way replies received. In addition to this the wireless Operator writes up in a Log book, which is called a 'Procès-verbal', and [*sic*] abbreviated account of the telegraphic work accomplished by his Station, and it is generally written up from hour to hour. This document is also, at the end of the voyage, sent to the Office which controls the station on the ship by the wireless Operator.[50]

Turnbull also detailed, in chronological order, six key ice messages with the replies thereto (if any) from the *Titanic*, and laid out the evidence for their transmission and receipt from service message forms submitted by the vessels concerned, backed up where possible by entries in the relevant procès-verbaux. The six messages may be summarised as follows:

— Message received by the *Titanic* from the SS *La Touraine*, steamer of the Compagnie Generale Trans-Atlantique, 12 April 1912, reporting crossing a thick ice field at Lat. 44.58 Long. 50.40, and that it saw another icefield and two icebergs Lat. 40.56 Long. 68.38. Reply from Captain Smith giving thanks. These were copies of official messages handed in to the Paris Company by the *La Touraine*.

— Message sent by the RMS *Caronia* to the *Titanic* 7.10 a.m. 14 April, 1912, stating that westbound steamers reported bergs, growlers and field-ice in 42 N from 49 to 51 W. Answered at 8.31 a.m. by Smith. There was a corroborative entry of outward message in the *Caronia*'s procès-verbal.

— Message sent by the German steamer SS *Amerika* of the Hamburg–Amerika Line to the Hydrographic Office, Washington 11.45 a.m. New York Time, 14 April 1912. The *Amerika* reported passing two large icebergs in 41.27 N 50.8 W on 14 of April. The message was sent to the *Titanic* for transmission via Cape Race to the Hydrographic Office. At the time of compilation it was not possible to prove it reached Cape Race as that station's returns were still awaited. There was proof in the *Amerika*'s procès-verbal that it was sent to the *Titanic*.

— Message from the *Baltic* to the *Titanic* sent 14 April 1912, at 11.52 a.m. NYT. This gave the information that the Greek steamer SS *Athinai* reported passing icebergs and a large quantity of field ice that day in Lat. 41.51 N. Long. 49.52 W. It continued: 'Last night we spoke German Oil Tank steamer Deutschland Stettin to Philadelphia not under control short of coal Lat. 40.42 N. Long. 55.11 W. Wishes to be reported to New York and other Steamers.' The service form reply at 12.55 p.m. confirmed receipt, and this was corroborated by the *Baltic*'s procès-verbal.

— Message from the *Californian* to the SS *Antillian* 14 April, sent 5.35 p.m. This gave the vessel's position as Lat. 42.3 N Long. 49.9 W, and reported three large bergs five miles southward. This was apparently received by the *Titanic* according to the evidence of the operator on the *Californian*. The communication between the *Californian* and the *Antillian* was confirmed by the *Californian* PV. It was noted that Bride admitted in the US Inquiry that he got a message from the *Californian*, but he stated it was from the *Californian* to the *Baltic*, not the *Antillian*. This was not confirmed by the *Baltic*'s records.
— Message from the *Mesaba* (of the Atlantic Transport Line) to the *Titanic* at 7.50 p.m. on 14 April. This ice report to the *Titanic* and all eastbound ships stated that in Lat. 42 N to 41.25 Long. 49 W to 50.30 W it saw much heavy pack ice and a great number of large icebergs, also field ice. The *Mesaba* operator had noted a reply of thanks, presumed to be from the *Titanic* but the note did not make this clear. The transmission was confirmed by the *Mesaba*'s procès-verbal.[51]

A letter from the Marine Company (again, probably written by Turnbull) to Guglielmo Marconi dated 13 June contains detailed notes for Marconi's assistance in submitting his own statement to the Inquiry. These illuminate the Company's activities, findings and contributions to the Inquiry to date. Thus far the Company had been asked for particulars of the Masters' Service Messages and the proving of them, together with information about the wireless installation on the *Titanic*, and the mode of procedure in sending, receiving and delivering telegrams. It had gathered together and had printed what seemed to be principal features of the Inquiry (so far as it concerned the Marine Company). These three documents were: the Masters' Service Messages sent to the *Titanic*, and the latter's acknowledgements of some of them; a substitute for the procès-verbaux of the *Titanic* and *Carpathia*, reconstituted as well as possible from the documents in the Company's possession, from 12 to 18 April; and information from the company's documents on what assistance was given by the *Asian* to a disabled tank steamer, which incident and reports thereon may have given rise to the report that the *Titanic* was being towed to Halifax.

Turnbull reported that all the *Titanic*'s records had been lost, but that the other stations' forms of the Masters' Service Messages sent to her, and the acknowledgements of them received from her, had been pasted in a book, which was now in possession of the court. The other messages sent and received by the *Titanic* from 12 April until she was lost had also been arranged in book form, and retained by the Company. Similarly the *Carpathia*'s messages had been arranged in four books: (1) messages sent and received by Mr Ismay; (2) messages sent and received by Captain Rostron; (3) private messages sent and received; (4) messages non-transmitted. Finally, another book of Masters' Service Messages had been compiled in anticipation of requirements from the records of ships in the vicinity of the

Titanic accident, advising one another of the presence of ice, between 11 and 16 April. All messages sent to the *Titanic* were in the two *Titanic* message books, one in the possession of the court, the other in the office.

Next, Turnbull turned to the relative significance of the various ice messages. Those of the *Empress of Britain* and *La Touraine* he considered of little importance as they reported ice well to the north of the *Titanic* wreck. Those of the *Caronia*, *Amerika*, *Baltic* and *Mesaba* were all-important as they warned of ice in a comparatively small parallelogram through which the *Titanic*'s track lay. The *Baltic* and *Mesaba* messages were looked upon as the most important – that of the *Baltic* especially, because of the acknowledgement sent by Smith. Referring to evidence given so far in the ongoing British Inquiry, Turnbull noted that Lord Mersey was not satisfied that the *Mesaba* message was ever delivered to the bridge, although he recognised that it was received by the *Titanic* operator. At this point Turnbull commented:

> Mr Phillips would receive the *Mesaba* message and it is inconceivable that he should not have delivered it at once. Sir Robert Finlay would like to make out that he was so busy sending private and fully-paid messages to Cape Race that he could not properly attend to messages that were not paid for. The records of Cape Race show, however, that there was a break in the communications at the time the *Mesaba* message was sent.[52]

This reference to Cape Race's procès-verbal is confirmed by the surviving record in the Marconi Archives: however, there is only a typescript summary of Cape Race's activities and it is unclear to what extent this was based on the original procès-verbal, which is not present. The summary states that the *Titanic* was in communication with Cape Race between 6.15 p.m. and 7.42 p.m. that evening, from which point there was a break until the *Titanic* resumed working with Cape Race at 9.20 p.m. *Mesaba*'s ice message had been transmitted at 7.50 p.m., eight minutes after the *Titanic* had cleared Cape Race.[53]

Referring to the refusal of Harold Bride to receive a Master's Service Message offered by the *Californian* – the fifth of the six key messages noted above – Turnbull noted that Bride, besides stating that he overheard the same message being delivered to another ship, said that he delivered the overheard message to the bridge, but that this point could not be proven. There was further controversy about a later message from the *Californian*, reporting that it was surrounded by ice and had stopped for the night. This message – only an hour or so before the iceberg was struck – was also refused, with the words 'keep out' as recorded in the evidence to the Inquiry of Cyril Evans, radio operator on the *Californian*. Evans stated the phrase was commonly used when an operator was working with another station, and was not taken as an insult.[54]

This account of the documents being compiled or assembled for the Inquiry is invaluable in placing the surviving documents in context, and also casts some light on the origin of certain records that are not among those in the Marconi Archives, notably the four *Carpathia* message books. These found their way on to the auction market in 1992 and were dispersed to museums and private owners.[55]

The substitute for the procès-verbaux of the *Titanic* and *Carpathia* mentioned by Turnbull in the foregoing letter drew on the procès-verbaux of some thirty-seven ship and shore stations to compile a record of the wireless activities of the operators on board the *Titanic* and *Carpathia*, whose records did not survive or were wanting. The two copies in the Marconi Archives show amendments and corrections, so may not be the final versions as submitted to the Inquiry.[56]

Likewise, four printed copies of the transcripts of Masters' Service Messages received by the *Titanic* survive in the collection.[57] One copy is annotated with references to the documents therein made in both the English (*sic*) and American inquiries. Once again, these feature the ice messages previously referred to, but with the addition of a message from the *Noordam* to the *Titanic* via the *Caronia*, together with a reply: this reported 'much ice in lat. 42.24 to 42.45, and long. 49.50 to 50.20'. The existence of this message was reported to Cunliffe in a letter from the Company dated 30 May 1912.[58]

It should be noted that while the series of procès-verbaux appears substantially complete, a few survive only in copy or summary form, such as for Cape Race, as mentioned above. The majority are the original forms written out in the wireless stations as the disaster unfolded, although at least some of those for the German Debeg Company appear to have been supplied as manuscript copies, as the serial numbers for different ships are consecutive. One procès-verbal is no longer present in the collection, although there is an original folder for it: that of the *Californian*.

INVESTIGATIONS INTO THE SINKING

An investigation into the sinking of the *Titanic* was begun by a subcommittee of the Committee of Commerce of the United States Senate astonishingly quickly – the day after the arrival of the *Carpathia* in New York. Cottam and Bride both gave evidence, as did Guglielmo Marconi, who had arrived in New York on the *Lusitania*. There was discussion of Bride's fitness to appear, with Captain Rostron commenting that both Bride's ankles and back were injured: despite this, he had been carried up to the Marconi operating room of the *Carpathia* to assist Cottam as much as he could.

Cottam was the first to appear, and he was questioned about the nature of working as the sole operator on a ship. He explained that he did not work fixed hours, but varied his routine according to need – for example, by working most of the time if close to New York. Cottam had been about to turn in at the time he received the *Titanic*'s distress call: he had only been waiting for confirmation of a message he had sent to the *Parisian*. While waiting, he had heard Cape Cod seeking to work with the *Titanic*, and he called the *Titanic* operator himself to let him know. The *Titanic* responded with a call for help. There was also considerable discussion about what

Cottam had been doing between leaving the site of the sinking and arriving in New York: the committee appeared under the impression that very little was going on during this period, this possibly based on the failure of so many to make contact with the *Carpathia*.[59]

When Harold Bride appeared, his evidence was notable for a vivid account of the final moments in the *Titanic*'s wireless cabin and of his own remarkable survival. He described how he and Phillips thought the lamps of their apparatus must be running down, as it was difficult to get a spark. Soon after this, they made preparations to leave the ship once the captain had given permission. Captain Smith did so about fifteen minutes before the ship disappeared, telling Phillips and Bride they had better look out for themselves. The two operators went out on deck, where Bride helped to bring down a collapsible lifeboat. A wave washed the boat over the side before it could be lowered properly, and took Bride with it too. He ended up underneath the upturned boat, he thought for about half or three-quarters of an hour. Eventually he managed to get on top of the boat, where there were already thirty to forty people. Phillips was also on the boat, but perished nevertheless. Those on the upturned boat were mostly men, and they remained on it until picked up by the *Carpathia*.[60]

At a later stage of the investigation, Harold Bride submitted to the subcommittee a written report he had made for the traffic manager of the Marconi Company, and this is transcribed in full in the evidence. Bride had compiled it because of the conflicting reports concerning the event, aiming to present as accurate a record as he could remember. His motivation appears to have been the manner and topics of cross-questioning at the investigation, and he particularly sought to explain and justify points that he had found difficult to convey. One such point was the reason for Phillips's annoyance when telling the *Frankfurt* operator to 'keep out': Bride considered Phillips was justified in sending this instruction during the emergency, as the *Frankfurt* appeared to be misunderstanding and interfering with communication with the *Carpathia*, which was on its way to give assistance.

Bride noted that he had kept the *Titanic*'s procès-verbal up to date, intending they should take a copy each when they left, but in the event there was no time to do so. Likewise, the immediate danger was pressing before it was necessary to make use of the emergency apparatus rather than the ship's dynamo.

On the *Carpathia*, Bride gave assistance to Harold Cottam after a short spell in the ship's hospital. He was told that a list of survivors had been sent to the *Minnewaska* and the *Olympic*. Hundreds of telegrams from survivors were waiting to go as soon as there was communication with shore stations. There was heavy traffic waiting for the *Carpathia* at the various shore stations; the two operators on the ship advised they would only accept service and urgent messages, and Bride understood that Captain Rostron had advised getting the survivors' traffic off first. Bride explicitly stated that neither he nor Cottam withheld news with the idea of making money – something on which

[Price 1s. 6d.]

In the Wreck Commissioners' Court.

SCOTTISH HALL,
BUCKINGHAM GATE,
Thursday, 2nd May, 1912.

PROCEEDINGS

BEFORE

THE RIGHT HON. LORD MERSEY,

WITH

REAR ADMIRAL THE HON. S. A. GOUGH-CALTHORPE, C.V.O., R.N.,
CAPTAIN A. W. CLARKE,
COMMANDER F. C. A. LYON, R.N.R.,
PROFESSOR J. H. BILES, LL.D., D.Sc.,
MR. E. C. CHASTON.

ON A FORMAL INVESTIGATION

ORDERED BY THE BOARD OF TRADE INTO THE

LOSS OF THE S.S. "TITANIC."

[*Transcribed from the Shorthand Notes of* W. B. GURNEY & SONS,
26, *Abingdon Street, Westminster, S.W.*]

FIRST DAY.

THE RIGHT HON. SIR RUFUS ISAACS, K.C., M.P. (Attorney-General), THE RIGHT HON. SIR J. SIMON, K.C., M.P. (Solicitor-General), MR. BUTLER ASPINALL, K.C., MR. S. A. T. ROWLATT and MR. RAYMOND ASQUITH (instructed by SIR R. ELLIS CUNLIFFE, Solicitor to the Board of Trade) appeared as Counsel on behalf of the Board of Trade.

THE RIGHT HON. SIR ROBERT FINLAY, K.C., M.P., MR. F. LAING, K.C., MR. MAURICE HILL, K.C., and MR. NORMAN RAEBURN (instructed by Messrs. Hill, Dickinson and Co.) appeared as Counsel on behalf of the White Star Line.

MR. THOMAS SCANLAN, M.P. (instructed by Mr. Smith, Solicitor) appeared as Counsel on behalf of the National Sailors' and Firemen's Union of Great Britain and Ireland, and of the personal representatives of several deceased members of the crew and of survivors who were members of the Union. Admitted on application —*See below.*

MR. BOTTERELL (instructed by Messrs. Botterell and Roche) appeared on behalf of the Chamber of Shipping of the United Kingdom. Admitted on application—*See below.*

MR. HAMAR GREENWOOD, M.P. (instructed by Messrs. Pritchard and Sons) watched proceedings on behalf of the Allan Line Steamship Company.

MR. HAMAR GREENWOOD, M.P. (instructed by Messrs. William A. Crump and Son) watched proceedings for the Canadian Pacific Railway Company.

Title page of the British Inquiry Proceedings, May 1912. (Oxford, Bodleian Library, MS. Marconi 291, p. 1.)

both he and Cottam had been pressed earlier. Bride was paid £500 by the *New York Times* for his story, with the approval of Guglielmo Marconi, who brought the reporter to see Bride.[61]

In the report on their investigation, the subcommittee explicitly stated that it did not believe the wireless operator on the *Carpathia* showed proper vigilance in handling the important work confided to his care after the accident.[62] This view does not appear to be borne out by the evidence relating to wireless communications to be found in the collection.

The British Inquiry into the sinking was held between May and July 1912, when witnesses had returned from the United States. Headed by Lord Mersey, the proceedings appeared more measured than the American investigation, although witnesses were cross-examined intensively. The Board of Trade had drawn up a series of questions to be answered, of which three were of particular relevance to wireless telegraphy.

Question 6 asked what installations for receiving and transmitting messages by wireless telegraphy were on board the *Titanic*; how many operators were employed in operating such installations; whether the installations were in good and effective working order, and the number of operators sufficient to enable messages to be received and transmitted continuously by day and night.

Question 9 sought to ascertain whether, after departure from Queenstown, information reached the *Titanic* by wireless messages (or otherwise by signals) of the existence of ice in certain latitudes. If so, what were such messages or signals and when were they received, and in what position or positions was the ice reported to be, and was the ice reported in or near the track actually being followed by the *Titanic*? Further, was her course altered in consequence of receiving such information, and, if so, in what way? And what replies to such messages or signals did the *Titanic* send, and at what times?

Question 18 queried what messages for assistance were sent by the *Titanic* after the casualty, and at what times; what messages were received by her in response, and at what times; by what vessels were the messages that were sent by the *Titanic* received, and from what vessels she received answers; what vessels other than the *Titanic* sent or received the messages at or shortly after the casualty in connection with such casualty; what were the vessels that sent or received such messages; were any vessels prevented from going to the assistance of the *Titanic* or her boats owing to messages received from the *Titanic* or owing to any erroneous messages being sent or received; and, in regard to such erroneous messages, from what vessels were they sent and by what vessels were they received and at what times, respectively.

Evidence was given on various days by several radio operators: besides Harold Bride and Harold Cottam, these were Cyril Evans of the *Californian*, J. Durrant of the *Mount Temple* and Stanley Adames of the *Mesaba*. In addition, G.E. Turnbull and Guglielmo Marconi appeared as witnesses. Adames appeared at a late stage, as he did not return to Britain until towards the end of the Inquiry.

Turnbull's evidence followed the lines as reported to Marconi by Turnbull in the correspondence above. Harold Bride once again went through his recollections of the days immediately before and after the sinking: there were some variations between his answers and the evidence presented to the US investigation, and these were picked up on cross-examination. Bride reiterated the problems experienced through the *Carpathia* and *Frankfurt* communicating with the *Titanic* at the same time during the emergency itself, and was cross-examined about the controversial non-acceptance of communications about ice from the *Californian* prior to the iceberg being struck. Evans also gave evidence about this, as noted by Turnbull above.

Cottam's evidence likewise reiterated much of what was said in America, recounting the moments when he picked up the *Titanic*'s distress signal almost by accident when about to go off-duty.

Guglielmo Marconi was questioned about the provision of marine wireless services in general, and the regulations governing emergency situations. Discussing the use of the CQD and SOS distress calls, he stated that CQD was to be signalled only on the order of the captain of the ship in distress, and in re-transmissions thereof. Marconi referred to the International Radio-Telegraph Convention of 1906, which laid down regulations including the use of SOS as the internationally recognised distress call from 1908; the CQD call had been used by Marconi operators before that and continued in use because of its familiarity. In the *Titanic* emergency both had been given out.

Guglielmo Marconi and Godfrey Isaacs (managing director of the Marconi Company) after the *Titanic* disaster, May 1912. (Oxford, Bodleian Library, MS. Photogr. c. 235, fol. 81.)

Report on the Loss of the "Titanic" (S.S.)

THE MERCHANT SHIPPING ACTS, 1854 to 1906.

IN THE MATTER OF the Formal Investigation held at the Scottish Hall, Buckingham Gate, Westminster, on the 2nd, 3rd, 7th, 8th, 9th, 10th, 14th, 15th, 16th, 17th, 20th, 21st, 22nd, 23rd and 24th May, the 4th, 5th, 6th, 7th, 10th, 11th, 12th, 13th, 14th, 17th, 18th, 19th, 21st, 24th, 25th, 26th, 27th, 28th and 29th June, at the Caxton Hall, Caxton Street, Westminster, on the 1st and 3rd July, and at the Scottish Hall, Buckingham Gate, Westminster, on the 30th July, 1912, before the Right Honourable Lord Mersey, Wreck Commissioner, assisted by Rear Admiral the Honourable S. A. Gough-Calthorpe, C.V.O., R.N.; Captain A. W. Clarke; Commander F. C. A. Lyon, R.N.R.; Professor J. H. Biles, D.Sc., LL.D., and Mr. E. C. Chaston, R.N.R., as Assessors, into the circumstances attending the loss of the steamship "Titanic," of Liverpool, and the loss of 1,490 lives in the North Atlantic Ocean, in lat. 41° 46′ N., long. 50° 14′ W. on the 15th April last.

REPORT OF THE COURT.

The Court, having carefully enquired into the circumstances of the above mentioned shipping casualty, finds, for the reasons appearing in the Annex hereto, that the loss of the said ship was due to collision with an iceberg, brought about by the excessive speed at which the ship was being navigated.

Dated this 30th day of July, 1912.

MERSEY,

Wreck Commissioner.

We concur in the above Report.

ARTHUR GOUGH-CALTHORPE.
A. W. CLARKE.
F. C. A. LYON.
J. H. BILES.
EDWARD C. CHASTON.

Assessors.

Wt. 4193.—292. 25.† 7/12. J. T. & S

A

Guglielmo Marconi's copy of the Report of the British Inquiry, July 1912. (Oxford, Bodleian Library, MS. Marconi 293, cover.)

Marconi also discussed the use of the Track and Communication charts used to facilitate communication between ships following the main shipping routes, such as the North Atlantic in this instance. One of the standard charts is preserved in the Marconi Archives, but it appears a special, larger-format version had been drawn up for use in the Court of Inquiry.

Several questions were put to Marconi about possible remedies for the difficulty in maintaining a continuous wireless watch on ships having only one operator. He was asked if a member of the crew who was not an expert could be trained to recognise a simple signal, keep watch while the qualified operator was off duty and call him in the event of an emergency. Marconi agreed that this was possible but thought that it would not be altogether reliable. He had an alternative to suggest: he thought it feasible to set up the wireless apparatus so that it caused a bell to ring upon receipt of a distress signal, and mentioned that he had been giving considerable thought to this possibility in the weeks since the accident. However, he emphasised that the best solution – especially as things stood – was to have two qualified operators to enable a continuous watch to be maintained.[63]

The report of the British Inquiry was remarkably low-key. The summary answers to the questions noted above were often the briefest of notes, although reference was made to detailed lists of the relevant wireless messages, which were included in the report. The resulting recommendations were perhaps of the greatest interest with regard to wireless provision, that most pertinent suggesting

> That on all such ships there should be an installation of wireless telegraphy, and that such installation should be worked with a sufficient number of trained operators to secure a continuous service by night and day. In this connection regard should be had to the resolutions of the International Conference on Wireless Telegraphy recently held under the presidency of Sir H. Babington Smith. That where practicable a silent chamber for 'receiving' messages should form part of the installation.[64]

THE LEGACY OF THE DISASTER

In the longer term, the inquiries and subsequent international negotiations resulted in the Convention for the Safety of Life at Sea in 1914. Sixteen nations were party to this, and it included a chapter on radiotelegraphy. This stipulated that all merchant ships of certain specifications and having fifty persons or more on board had to have a radiotelegraph installation, with a range of at least one hundred sea miles by day. A continuous watch had to be maintained during navigation by one or more operators, assisted if necessary by one or more certificated watchers. However, it included the proviso that if efficient automatic calling apparatus were to be invented, then a continuous watch might be maintained by this means with the agreement of the appropriate regulators.[65]

In the event, a means of automatic calling was devised but was not introduced until after the First World War. On a slightly different note, in the year following the sinking an ice patrol was introduced in the North Atlantic shipping lanes. Vessels dedicated to keeping a watch for ice, and conveying warnings by wireless, maintained a continuous reconnaissance for potential danger. The success of this was such that the incidence of collisions with icebergs was reduced to nil.[66]

It may be seen that wireless telegraphy had undergone considerable advances in the years from its inception to the occurrence of the *Titanic* disaster, such that it was able to play a crucial role in obtaining assistance for the stricken vessel. It was also instrumental in conveying news of the sinking and the notification of survivors to the wider world. However, the difficulties in communicating with certain vessels that might have come to the aid of the *Titanic*, and the confusion surrounding communications in the aftermath of the sinking – not least due to interference because of the number of stations seeking to make contact – showed up the inadequacies of the contemporary arrangements, and led to significant improvements in future. Wireless telegraphy was not quite fully fledged at the time of the sinking, and the disaster played a significant part in taking its development (both technical and operational) a stage further.

RECOGNITION OF THE ROLE OF THE WIRELESS OPERATORS

It is clear from the foregoing that the technical story of wireless telegraphy in the *Titanic* disaster is constantly overlain by the human and personal side of the tragedy as it unfolded. The wireless record depicts the confidence of the *Titanic*, its crew and its passengers as they began the maiden voyage, which quickly turned to panic and confusion when the iceberg was struck. The poignant messages for survivors bring home the personal loss suffered by many, and the huge efforts of the wireless operators under these conditions were truly heroic.

On 15 April, a telegram was sent by G.A. Phillips to the Marconi Company:

> Any tidings of Titanic.[67]

N.B.—This Form must accompany any inquiry respecting this Telegram.

POST OFFICE TELEGRAPHS.

If the Receiver of an Inland Telegram doubts its accuracy, he may have it repeated on payment of half the amount originally paid for its transmission, any fraction of 1d. less than ½d. being reckoned as ½d.; and if it be found that there was any inaccuracy, the amount paid for repetition will be refunded. Special conditions are applicable to the repetition of Foreign Telegrams.

Office of Origin and Service Instructions. Farncombe

Charges to pay s. d.

Office Stamp.

Handed in at .M., Received here at

TO Reply Paid Marconi Co
7 York Buildings Ldn WC

Any tidings of Titanic
Phillips Farncombe

RECEIVED 15 APR 1912
HANDED TO
ATTENDED TO
BY
DATE

Telegram from G.A. Phillips, father of Jack, to the Marconi Company enquiring about news of the *Titanic*. (Oxford, Bodleian Library, MS. Marconi 256, fol. 66.)

11 Farncombe Rd
Godalming
8 September 1912

Dear Sir

I am exceedingly obliged to you for sending the enclosed papers for my perusal.

It is very gratifying to us that such widespread appreciation of my son's conduct, and sympathy has been shewn.

I have taken the liberty of copying these papers, as it will be a source of comfort to us in the future, in reminding us that our son's sacrifice was not in vain

Yours faithfully
G. A. Phillips

W. W. Bradfield Esq.

Letter from G.A. Phillips to W.W. Bradfield, 8 September 1912, acknowledging the many messages in appreciation of his son Jack. (Oxford, Bodleian Library, MS. Marconi 256, fol. 141.)

This was from Jack Phillips's father, who had presumably heard reports of the sinking and sought news of his son. The company replied briefly on the same day:

> Have no information direct from the Titanic.[68]

In due course formal condolences were sent to G.A. Phillips and his wife mourning the loss of Jack Phillips in the sinking. The company received numerous messages of appreciation of the devotion to duty of both Phillips and Bride. These sentiments, mostly from postal and telegraphic companies and unions at home and abroad, were conveyed to Phillips's family, and it is evident from correspondence that they were much appreciated.[69] Phillips's contribution was further recognised by a substantial memorial to him erected in his home town of Godalming, and this stands as a lasting commemoration of the commitment and bravery of Phillips and other operators who did their utmost to assist others in the most difficult of circumstances.

PROCÈS-VERBAUX

The PVs or procès-verbaux are logbooks kept by the wireless operators as a record of all they sent and received, and sometimes messages which they overheard passing between other ships. As such they provide by far the best record of events in the North Atlantic on the night of 14 April 1912. The *Titanic*'s PV did not survive, in spite of Harold Bride having made an additional copy to take with him, but the PVs of other ships within range of its powerful wireless equipment enable an almost complete picture to be built up of the messages being sent from the stricken ship.[1] The following pages contain extracts from the PVs of twelve ships and two shore stations. These document both communications with the *Titanic* during the disaster, and the subsequent difficulties and confusion involved in transmitting news of the sinking and lists of survivors from the *Carpathia* to the shore. The *Carpathia*'s PV is, unfortunately, a later reconstruction. The original simply reads 'Rescue of *Titanic*. Apparently too busy to keep PV going.' The document submitted to the Inquiries (and printed here in full) is based in part upon Harold Cottam's recollection of messages sent and received, and in part upon information relating to the

Carpathia in the PVs of other ships and shore stations. These twelve ships' PVs include almost every message received by or sent to the *Titanic* from the first distress call to the final, broken message picked up by the *Virginian*.

The two shore station PVs (those of Sable Island, off the coast of Nova Scotia, and Siasconset, at the eastern point of Nantucket Island, Massachusetts) were selected as offering the best representation of the messages (or sometimes lack of messages) reaching the shore from the *Carpathia*. The *Carpathia*'s wireless equipment had only a very short range, so the majority of messages had to be relayed via ships which were nearer the shore. This process was hampered by the communications of other ships, many of which had more powerful transmitters and simply drowned out ('jammed') the *Carpathia*. In addition to this, Harold Cottam was instructed by Captain Rostron to prioritise official messages, the names of survivors, and personal messages from the survivors.[2] He therefore consistently refused to take messages from the shore throughout the *Carpathia*'s journey to New York. Cottam was also exhausted, working ceaselessly for several days and nights, only relieved for short periods by Harold Bride, who had to be carried up to the wireless room to assist Cottam, having temporarily lost the use of his legs as a result of his ordeal.[3] A combination of these factors resulted in many enquiries from the shore going unanswered, and the frustration this caused can be clearly seen in the shore station PVs.

In a number of places in the PVs, sections which were not relevant to the *Titanic* and its story have been omitted. The only exception to this is the PV of Siasconset, where long periods between messages from the *Carpathia* have been retained for the insight they offer into the volume of traffic being handled by the operators and the inherent difficulties of their role.

PROCES-VERBAL. S.S. MOUNT TEMPLE 32

TIME.	REMARKS.	Call Letters.	Duration.	Condition.
p.m.	Saturday April 13th 1912 (Continued)			
7.50	Sigs M.M.V. – Nil			
8.18	Ex T.Rs Corinthian – East – Nil.	M.K.N		G
8.30	Ex T.Rs Athenai – West – Nil	M.T.S.		G
8.32	Sigs M.P.A. – Nil.			
8.45	Clear 1 from M.K.N.			
9.15	" 1 to "			
10.35	Sigs with M.K.N.			
a.m.	Sunday April 14th 1912			
6.35	Ex T.Rs Californian – West – Nil			
6.50	Ex T.Rs Parisian – West – Nil			
7.45	Titanic Working			
9.53	Ex T.Rs Amerika – East – Nil.			
10.10 till	Ex T.Rs Baltic – East – Nil			
11.15	Standing By – Several Ships Busy.			
11.18	Sigs M.G.N. – Nil			
11.37	Ex T.Rs Carpathia – East – Nil.			
11.38 till	Ex T.Rs Antillian – East – Nil			
8.30	M.G.Y. and M.C.E. Working			
9.55	Sigs M.P.A. – Nil			
10.25	Titanic Sending C.Q.D. Says Requires … Gives Position – Cannot Hear Me. A… Captain – His Position 41.46.N. 50.14 W. …			
10.35	M.G.Y. Gets M.P.A. and Says Struc… Iceberg Come To Our Assistance At …			
10.40	M.G.Y. still Clg C.Q.D.			
10.40	Our Captain Reverses ~~Engines~~ Ship {Hea… 50…			
10.48	Frankfurt gives M.G.Y his position…			
10.55	M.G.Y. Clg S.O.S. (39.47.N. – 52…			
10.57	" " M.K.C.			
10.59	" wkg M.P.A.			

33

THE MARCONI INTERNATIONAL MARINE COMMUNICATION CO., LTD.

PROCES-VERBAL. S.S. MOUNT TEMPLE

TIME.	REMARKS.	Call Letters.	Duration.	Condition.
p.m.	Sunday April 14th 1912 (Continued)			
11.0	M.G.Y. Calling M.G.N. and C.Q.D.			
11.10	" " C.Q.D.			
11.20	" tells M.K.C. "Captain says get your boats ready." "Going down fast at the head"			
11.25	D.F.T. says "Our Captain will go for you"			
11.27	M.G.Y. calls M.B.C. – Ser			
11.30	" " C.Q.D.			
11.35	M.K.C sends M.S.G to M.G.Y. – M.G.Y. replies "We are putting the women off in the boats" –			
11.41	M.G.Y says C.Q.D. Engine Room Flooded.			
11.43	Tells M.K.C. Sea Calm.			
11.45	D.F.T. Asks "Are there any boats around you already?" No Reply.			
11.47	M.K.C sends M.S.G to M.G.Y. – He sends Rd.			
11.55	S.B.A. and D.F.T. Clg M.G.Y. – No Reply.			
a.m.	Monday April 15th 1912.			
12.10	M.K.C D.F.T. & M.K.C Calling M.G.Y – No Reply.			
12.25	S.B.A. tells D.F.T. He is 70 Miles from M.G.Y.			
12.50	All Quiet Now – M.G.Y hasn't spoken since 11.47 p.m.			
1.25	M.P.A sends "If you are there, we are firing rockets."			
1.40	M.P.A. Clg M.G.Y.			
1.58	S.B.A Thinks He Hears M.G.Y. So Sends – "Steaming Full Speed To You; shall arrive you six in morning Hope You Are Safe. We are only 50 Miles Now"			
2.0	M.P.A Clg M.G.Y			
3.0	All quiet. – We're Stopped Amongst Pack Ice,			

Pages from the *Mount Temple*'s procès-verbal. (Oxford, Bodleian Library, MS. Marconi 261, fols 208–9.)

S.S. *MOUNT TEMPLE*, PROCÈS-VERBAL

NYT	**Sunday 14 April 1912**[4]
10.25 pm	Titanic [MGY] sending CQD Says requires assistance Gives position – cannot hear me Advise my captain – His position 41.46 N. 50.14 W (Nobody else answers).
10.35 pm	MGY [Titanic] gets MPA [Carpathia] and says: Struck iceberg Come to our assistance at once.
10.40 pm	MGY [Titanic] still clg CQD.
10.40 pm	Our captain reverses ship (We are about 50 miles off).
10.48 pm	Frankfurt [DFT] gives MGY [Titanic] his position (39.47 N–52.10 W).
10.55 pm	MGY [Titanic] clg SOS.
10.57 pm	MGY [Titanic] clg MKC [Olympic].
10.59 pm	MGY [Titanic] clg MPA [Carpathia].
11.00 pm	MGY [Titanic] calling MGN [Virginian] and CQD.
11.10 pm	MGY [Titanic] clg CQD.
11.20 pm	MGY [Titanic] tells MKC [Olympic] – "Captain says get your boats ready" "Going down fast at the head".
11.25 pm	DFT [Frankfurt] says "Our Captain will go for you".
11.25 pm	MGY [Titanic] calls MBC [Baltic].
11.30 pm	MGY [Titanic] calls CQD.
11.35 pm	MKC [Olympic] sends MSG to MGY [Titanic] – MGY [Titanic] replies "We are putting the women off in the boats".
11.41 pm	MGY [Titanic] says CQD Engine room flooded.
11.43 pm	Tells MKC [Olympic] Sea calm.

11.45 pm	DFT [Frankfurt] asks "Are there any boats around you already?" No reply.
11.47 pm	MKC [Olympic] sends MSG to MGY [Titanic] – He sends Rd.
11.55 pm	SBA [Birma] and DFT [Frankfurt] clg MGY [Titanic] – No Reply.
	Monday 15 April 1912
12.10 am	MKC [Olympic], DFT [Frankfurt] & MBC [Baltic] calling MGY [Titanic] – No reply.
12.25 am	SBA [Birma] tells DFT [Frankfurt] he is 70 Miles from MGY [Titanic].
12.50 am	All quiet now – MGY [Titanic] hasn't spoken since 11.47pm.
1.25 am	MPA [Carpathia] sends "If you are there, we are firing rockets."
1.40 am	MPA [Carpathia] clg MGY [Titanic].
1.58 am	SBA [Birma] thinks he hears MGY [Titanic] so sends – "Steaming full speed to you; shall arrive you six in the morning Hope you are safe We are only 50 miles now".
2.00 am	MPA [Carpathia] clg MGY [Titanic].
3.00 am	All quiet – We're stopped amongst pack ice.
3.05 am	SBA [Birma] and DFT [Frankfurt] working.
3.20 am	SBA [Birma] and DFT [Frankfurt] working.
3.25 am	We back out of ice & cruise around – big bergs about.
3.25 am	Answer MWL [Californian] CQ – Nil (Advise him of MGY [Titanic] and send him MGY [Titanic]'s position).
3.40 am	MWL [Californian] wkg DFT [Frankfurt] – DFT [Frankfurt] does the same.
4.00 am	MWL [Californian] wkg MGN [Virginian].
4.45 am	MWL [Californian] wkg SBA [Birma].
5.20 am	Sigs MWL [Californian] – wants my position – send it.
6.00 am	Much jamming MPA [Carpathia] & MWL [Californian] in sight.
6.45 am	MPA [Carpathia] reports rescued 20 boatloads.

7.15 am	More jamming – MPA [Carpathia] working MBC [Baltic].
7.30 am	MBC [Baltic] sends MSG to MWL [Californian].
7.40 am	MPA [Carpathia] calls CQ and says "No need to std bi him" Advise my Captain who has been cruising round the ice field with no result Ship reversed.
	MPA [Carpathia] and MKC [Olympic] very busy all day – much jamming going on.
9.15 am	MPA [Carpathia] and MKC [Olympic] still busy.

S.S. *VIRGINIAN*, PROCÈS-VERBAL

NYT	**Sunday 14 April 1912**[5]
9.45 pm	Standing by MCE [Cape Race] working continuously with MGY [Titanic].
10.00 pm	Bi for Cape Cod [MCC].
11.00 pm	MCC [Cape Cod]'s sigs scarcely audible Unable to read a single word Press on hard.
11.10 pm	Hear MGY [Titanic] calling CQ and giving his position as 41.46 N 50.14 W.
11.12 pm	Call MGY [Titanic] but get no response MCE [Cape Race] then calls me and asked me report to Capt that Titanic struck iceberg and requires immediate assistance.
11.30 pm	I to MCE [Cape Race] (MSG) MCE [Cape Race] informs MGY [Titanic] that we are going to his assistance Our position 170 miles North of MGY [Titanic].
11.35 pm	"MKC [Olympic] de [from] MGY [Titanic] —— Sinking We are putting passengers and —— off in small boats —— weather clear MGY [Titanic]."
11.50 pm	Cape Race [MCE] says:–"MGN [Virginian] pse tell ur Captn this – The MKC [Olympic] is making all speed for MGY [Titanic] but his position is 40.32 N 61.18 W You are much nearer to him He is already putting women off in boats and he says weather is clear and calm MKC [Olympic] is the only one we have heard say is going to his assistance The others must be long way from MGY [Titanic]."

Monday 15 April 1912

12.05 am	Cape Race [MCE] says "We have not heard MGY [Titanic] for about half an hour His power may be gone."
12.10 am	Hear MGY [Titanic] calling very faintly His power greatly reduced.
12.20 am	Hear two Vs signalled faintly in spark similar to MGY [Titanic]'s Probably adjusting spark.
12.27 am	MGY [Titanic] calls CQ Unable to make out his signal Ended very abruptly as if power suddenly switched off His spark rather blurred or ragged Called MGY [Titanic] & suggested he shd try emergency set but heard no response.
12.45 am	1 from MCE [Cape Race] (MSG).
1.15 am	Exchange sigs Baltic [MBC] He tries to send us MSG for MGY [Titanic] but his sigs died utterly away.
2.15 am	Sigs Russian American Liner "Birma" [SBA] says he is 55 miles from Titanic but cannot hear anything of him.
8.30 am	Standing bi (both operators) all night during which we exchanged sigs and compared positions with Californian – Frankfurt – Parisian and Carpathia all going to assistance of MGY [Titanic]. At 8 am Carpathia [MPA] said:– "Tell your Skipper we are leaving here with all on board – about 800 passengers Everything OK Please return to your northern course."
8.35 am	1 to Parisian (MSG).
8.40 am	Trs Iroquois [MEI] & Navahoe [MEN].
9.20 am	2 to Baltic [MBC].
3.00 pm	Stdg bi for MPA [Carpathia] all day Have offered him MSG sevl times He now refuses to take it and refuses to give our Captn any information whatever with respect to Titanic. Exchanged sigs with Hudson [MHS] and Olympic [MKC].
4.00 pm	Standing bi.
4.30 pm	Sigs MKC [Olympic] nil.

PROCES-VERBAL. S.S. "VIRGINIAN"

TIME.	REMARKS.	Call Letters.	Duration.	Condition.
	with Californian – Frankfurt – Parisian and Carpathia all going to assistance of	MGY		
	At 8am Carpathia said – "Tell your Skipper we are leaving here with all on board – about 800 passengers. Everything O.K. Please return to your northern course."			
8.35	1 to Parisian (MGY)			
8.40p	Trs Iroquois & Navahoe			
9.20	2 to Baltic			
3pm	Stdg bi for MPA all day. offered him MGY msg sevl times. He refuses to take it and refuses our Captn any informatn whatever respect to Titanic. Exchanged sigs with Hudson an			
4pm	Standing bi.			
4.30	Sigs MKC nil			
"	Msgs Bohemian nil board			
6.30p	1 from MEL, 1 to him.			
7.50p	Trs Macedonia			
8.15p	Sigs MEN. 8/25p Sigs MEL			
8.50p	Trs Ypiranga			
9.20p	Trs Prinz Adalbert			
10.0p	Bi for MGB. 10/30 press f first received greater part K's strong.			
	Standing bi. MKC wkg MCE with We have no tfc so deem it advisa to obviate risk of jmg MKC			

142 309

THE MARCONI INTERNATIONAL MARINE COMMUNICATION CO., LTD.

PROCES-VERBAL. S.S. "VIRGINIAN" 142

TIME.	REMARKS.	Call Letters.	Duration.	Condition.
	Sunday 14th April 1912			
11.35	"MKC de MGY — Sinking. We are putting passengers and — off in small boats — weather clear. MGY."			
11.50	Cape Race says: "MGN pse tell yr Captn this – The MKC is making all speed for MGY but his position is 40.32 N 61.18 W You are much nearer to him. He is already putting women off in boats and he says weather is clear and calm. MKC is the only one we have heard say is going to his assistance. The others must be long way from MGY."			
	Monday 15th April 1912			
12.50a	Cape Race says "We have not heard MGY for about half an hour. His power may be gone."			
12.10	Hear MGY calling very faintly his power greatly reduced.			
12.20	Hear two 'V's signalled faintly in spark similar to MGY's. probably adjusting spark.			
12.27	MGY calls CQ. Unable make out his signal. Ended very abruptly as if power suddenly switched off. His spark rather blurred or ragged. Called MGY [illegible] but heard no response.			
12.45	1 from MCE (MGY)	MCE		Good.
1.15	Exchange sigs Baltic. He tries send us msg for MGY but his sigs died utterly away.			
2.15	Sigs Russian American Liner Birma says he is 55 miles from Titanic but cannot hear anything of him.			

Pages from the *Virginian*'s procès-verbal. (Oxford, Bodleian Library, MS. Marconi 261, fols 309–10.)

4.30 pm	TRs Bohemian [MEL] nil bound west.
6.30 pm	I from MEL [Bohemian], I to him.
7.50 pm	Trs Macedonia [MDT].
8.15 pm	Sigs MEN [Navahoe].
8.25 pm	Sigs MEL [Bohemian].
8.50 pm	Trs Ypiranga [DYA].
9.20 pm	Trs Prinz Adalbert [DDZ].
10.00 pm	Bi for CC [Cape Cod].
10.30 pm	Press finished first Received greater part of it X's strong.
	Standing bi MKC [Olympic] wkg MCE [Cape Race] with heavy traffic We have no tfc so deem it advisable to stand bi and so obviate risk of jmg MKC [Olympic].

R.M.S. *OLYMPIC*, PROCÈS-VERBAL

NYT	**Sunday 14 April 1912**[6]
10.50 pm	Hear MGY [Titanic] signalling to some ship, and saying something about striking ice-berg, not sure if is MGY [Titanic] who has struck an ice berg, I am interfered by X's and many stns wkg.
11.00 pm	Hear MGY [Titanic] calling CQD Answered MGY [Titanic].
11.10 pm	MGY [Titanic] replies and gives me his position 41.46N. 50.14W and says we have struck an ice berg Reported same to Bridge immediately, our distance from MGY [Titanic] 505 miles.
11.20 pm	Sigs with MGY [Titanic] He says tell Captain get your boats ready and what is your position.
11.35 pm	Sent message to MGY [Titanic] giving our position.

Pages from the *Olympic*'s procès-verbal. (Oxford, Bodleian Library, MS. Marconi 261, fols 230–31.)

230

43

THE MARCONI INTERNATIONAL MARINE COMMUNICATION CO., LTD.

PROCES-VERBAL. S.S. OLYMPIC

TIME.	REMARKS.	Call Letters.	Duration.	Condition.
	~~Monday~~ Sunday 14th Apl (contd)			
3.15	Ica way MSB			
5.20	Tfc with MSB at ints.			
7.35	trying work MSB x's coming on bad + SD weak.			
.30	Calling MZD with message, unable read him thro X's			
.50	Sigs MGU.			
.15	Std bi for CC.			
.45	Recd 4 msgs from CC.			
.50	Hear MGY signalling to some ship, and saying something about striking ice-berg, not sure if it is MGY who has struck an ice berg, I am interfered by X's and many stns wkg.			
.0	Hear MGY calling CQD. answered MGY.			
.10	MGY replies and gives me his position 41.46 N. 50.14 W and says we have struck an ice berg. Reported same to Bridge immediately, our distance from MGY 505 miles.			
.20	Sigs with MGY he says tell Captain get your boats ready and what is your position.			
.35	Sent message to MGY			

PROCES-VERBAL. S.S. OLYMPIC

TIME.	REMARKS.	Call Letters.	Duration.	Condition.
	Sunday 14th continued.			
11.40p	MGY says tell captain we are putting the passengers off in small boats.			
11.45	Asked MGY what weather [he] had. he says clear & calm.			
	[...]t message to MGY [...]ing lighting up [...] possible boilers as fast [...] can.			
	[...]D calling me with tfc [...] him std bi for a [...]le having urgent [...]munication with MGY.			
	Monday 15th Apl			
	[...] Hellig Olav. Nil. DHO			G
	[...] sigs strong.			
	[...]d him if he had [...]d anything about MGY [...] says no.			
	[...]ping strict watch [...]s hear nothing more [...]m MGY.			
	[...]ing MZD at intervals. [...] reply from him.			
	[...]tc with MSD. now [...]ht. Ask MCE via MSD what news of	MGY.		
	[...]a Bretagne sends ice MLB			G
	[...] same as the "G. Washington's"			

11.40 pm	MGY [Titanic] says tell Captain we are putting the passengers off in small boats.
11.45 pm	Asked MGY [Titanic] what weather he had He says clear & calm.
11.50 pm	Sent message to MGY [Titanic] saying lighting up all possible boilers as fast as can.
11.55 pm	MSD [Sable Island] calling me with tfc Told him std bi for a while having urgent communication with MGY [Titanic].
	Monday 15 April 1912
12.30 am	TR Hellig Olav [DHO] Nil His sigs strong Asked him if he had heard anything about MGY [Titanic] He says no Keeping strict watch but hear nothing more from MGY [Titanic]. Calling MSD [Sable Island] at intervals No reply from him.
4.15–5.20 am	Tfc with MSD [Sable Island] Now daylight Ask MCE [Cape Race] via MSD [Sable Island] what news of MGY [Titanic].
5.30 am	TR La Bretagne [MLB] sends ice report Same as the "G. Washington's".
5.40 am	TR Asian [MKL] with German oil tank in tow for Halifax Asked what news of MGY [Titanic]? Sends service later saying "heard MGY [Titanic] v faint wkg C. Race up to 10 pm Local Time Finished callg SOS midnight.
7.00 am	205 [miles] SSE of MSD [Sable Island] nil.
7.10 am	Sigs MKL [Asian].
7.40 am	Tfc with MSD [Sable Island] SG fm MCE [Cape Race] via SD [Sable Island] saying CE [Cape Race] can read us & tells us to tune for him Call MCE [Cape Race] but cannot hear him at all.
7.50 am	1 to MKL [Asian].
8.05 am	TR Athinai [MTI] Nothing new of MGY [Titanic].
8.05 am	TR Scandinavian [MNC] bnd East.
8.30 am	2 fm MKL [Asian].
9.25 am	TR Parisian [MZN] Says "had worked MGY [Titanic] 8.30 l[ast] night & GY [Titanic] was wkg CE [Cape Race] when MZN [Parisian] turned in at 11.15 ship's reading Information recd Carpathia [MPA] has picked up 20 boats with psgrs –Baltic [MBC]

returning to MGY [Titanic] Don't know if latter sunk MWL [Californian] abt 50 miles astern of us."

10.10 am	Tfc with DKB [Berlin].
10.12 am	TR Mesaba [MMV] R'd ice report.
10.30 am	Tfc with MSD [Sable Island] & MZN [Parisian].
10.55 am	TR C. Race [MCE] He just audible 350 miles Knows nothing more of MGY [Titanic].
Noon	With much timely assistance from MNC [Scandinavian] cleared 3 to MCE [Cape Race] & r'd five mges Bi for MNC [Scandinavian]'s lunch.
12.25 pm	MSG to MZN [Parisian].
12.50 pm	Recd long ice report from MZN [Parisian].
1.25 pm	Trying to receive from MCE [Cape Race] His sigs dead weak, and DKB [Berlin] is jamming me badly Told DKB [Berlin] to be careful and not jam me MNC [Scandinavian] assists me to receive from MCE [Cape Race].
1.40 pm	Recd 1 from MCE [Cape Race]. I then informed MCE [Cape Race] that it was no use sending me messages from newspapers asking us to send news of Titanic as we had no news and told him to std bi as I must get hold of some ship who has news of MGY [Titanic]. Called CQ to std bi.
2.00 pm	Established comm. with Carpathia [MPA] his sigs good He gives me outline of what has happened and I report same to Commander immediately.
2.35 pm	Sent MSG to MPA [Carpathia].
2.40 pm	Sigs MGN [Virginian].
3.15 pm	MPA [Carpathia] gives me his 3 P and several notes for Capt.
3.35 pm	Sent MSG to MPA [Carpathia] asking for particulars of MGY [Titanic] so that we can report immediately via MCE [Cape Race].
4.00 pm	MPA [Carpathia] sends short account of disaster to Captain.
4.15 pm	Sigs MPA [Carpathia].

4.40 pm	2 msgs to MCE [Cape Race].
4.50 pm	1 to MPA [Carpathia] Sigs MWL [Californian].
5.20 pm	Ice report from MWL [Californian] who is 200 miles out of his course.
5.45 pm	MPA [Carpathia] starts sending thro names of survivors aboard.
7.35 pm	Cleared 322 first & 2nd class passengers names Sent 5 to him.
7.50 pm	Sigs MCE [Cape Race] who has 30 mges for us He v weak and interfered with by X's.
8.00 pm	Trying to read MCE [Cape Race], his sigs weak, impossible read him.
8.25 pm	Recd 5 MSGs from MPA [Carpathia] Asked him if he had list of 3rd class & crew survivors ready, he says no, will send them soon.
8.35 pm	Sent 1 to MWL [Californian].
8.45 pm	Recd 1 from MWL [Californian].
	Std bi for MPA [Carpathia], calling him frequently, nothing further from him. Fear no chance now of receiving remaining list of survivors.
10.00 pm	Started calling MCE [Cape Race] with list of 1st & 2nd class survivors from Titanic, unable raise him.
10.30 pm	MSD [Sable Island] answers me and offers tfc, told him have list of survivors here please take them. MSD [Sable Island] gives G and I commenced sending list.
	Tuesday 16 April 1912
12.20 am	MCE [Cape Race] comes in sigs good, says he can read me OK, and he has already recd names I have been sending to MSD [Sable Island], as MCE [Cape Race] is strong and I have great difficulty in reading MSD [Sable Island] thro X's, I send the remaining names to MCE [Cape Race].
2.30 am	Completed sending list of survivors to MCE [Cape Race] and then send MSG's which I received from MPA [Carpathia]. Commenced receiving tfc from MCE [Cape Race] distance 370 [miles] SE.

3.10 am	4 messages in all rd fm MCE [Cape Race] then his sigs gradually die away during fifth mge Daylight breaking here Sent several times to CE [Cape Race] that can't work him now.
3.15 am	United Wireless S.S. Marengo [UMO] rpts Virginian [MGN] had been to rescue but when nearly there had turned back on course.
3.35 am	TR Virginian [MGN] Confirms above.

R.M.S. *OLYMPIC*, CAPTAIN'S REPORT ON WIRELESS OPERATORS

20th April 1912[7]

CONFIDENTIAL REPORT

To: The Marconi International Marine Communication Company, Ltd.,
York Buildings, Adelphi, London, WC.

The following is a Report on Mr. C.J. Moore & A. Boyd
Marconi Operator on board this Vessel for voyage ending 20 [April] 12

Attention to Duties	VG
General Conduct	VG
Personal appearance	VG
Sobriety	VG

Remarks Can scarcely speak too highly of these Officers' conduct at a very trying & anxious time. Mr. Moore, especially, worthy of promotion. Thoughtful & reliable.

[signed] Jas Haddock Commander.

R.M.S. *CARPATHIA*, RECONSTRUCTED PROCÈS-VERBAL

From 5.10 p.m. Sunday 14th April, at which time she first established communication with the S/S "TITANIC", until she arrived at New York with the survivors on board (9.45 p.m. Thursday 18th April).

NYT	**Sunday 14 April 1912**[8]
5.10 pm	Trs with S.S. "Titanic" bound west. One "S" message received.
5.30 pm	Signals exchanged with the "Titanic" at frequent intervals until 9.45 p.m.
10.00 pm	Goodnight to "Mount Temple", his sigs very weak. [Original PV ends here with note: 'Rescue of Titanic Apparently too busy to keep PV going'.]
11.20 pm	Heard "Titanic" calling "SOS" and "CQD" Answered him immediately "Titanic" says "Struck iceberg, come to our assistance at once, position Lat. 41.46N. Long. 50.14W." Informed bridge at once.
11.30 pm	Course altered, proceeding to the scene of the disaster.
11.45 pm	"Olympic" working to "Titanic" "Titanic" says "Weather is clear and calm Engine room getting flooded".
	Monday 15 April 1912
12.10 am	"Titanic" calling "CQD" His power appears to be greatly reduced.
12.20 am	"Titanic" apparently adjusting spark gap, he sending "Vs" Signals very broken.
12.25 am	Calling "Titanic" No response.
12.28 am	"Titanic" calls "CQD" His signals blurred and end abruptly.
12.30 am	Calling "Titanic" at frequent intervals; keeping close watch for him, but nothing further heard.
1.25 am	Called "Titanic" and told him we are firing rockets; no signs of any response.
1.30 am	Continue to call "Titanic" at frequent intervals, but without success.
Daybreak	"CARPATHIA" ARRIVES ON THE SCENE OF THE DISASTER.
5.05 am	Signals with "Baltic", but unable to read him owing to continual disturbance atmospherics, etc.

6.45 am	Signals with "Mount Temple"; informed him we were now rescuing "Titanic's" passengers.
7.07 am	Received following message from "Baltic":– "TO CAPTAIN CARPATHIA:–CAN I BE OF ANY ASSISTANCE TO YOU AS REGARDS TAKING SOME OF THE PASSENGERS FROM YOU. WILL BE IN THE POSITION ABOUT FOUR THIRTY. LET ME KNOW IF YOU ALTER YOUR POSITION. COMMANDER "BALTIC".
7.10 am	Sent following reply to "Baltic":–"AM PROCEEDING FOR HALIFAX OR NEW YORK FULL SPEED, YOU HAD BETTER PROCEED TO LIVERPOOL. HAVE ABOUT 800 PASSENGERS ABOARD".
7.40 am	Advised "Mount Temple" to return to his course, as there is no further need for you to stand by; nothing more can be done We have rescued 20 boat loads of the "Titanic's" passengers.
8.00 am	Advised "VIRGINIAN":– "WE ARE LEAVING HERE WITH ALL ON BOARD – ABOUT 800 PASSENGERS – PLEASE RETURN TO YOUR NORTHERN COURSE".
10.00 am	Signals with "Californian".
2.00 pm	Trs with "Olympic".
2.10 pm	Sent news of the disaster to "Olympic", saying we had rescued about 670 passengers.
2.35 pm	Following message received from "Olympic":– "7.12 pm G.M.T. Position 41.17N, 53.53W Shall I meet you and where? Steering east true HADDOCK".
3.15 pm	Replied to "Olympic" 7.30 p.m. G.M.T. "Carpathia" 41.15N., 51.45W., Am steering South 87 West true returning to New York with "Titanic's" passengers ROSTRON.
3.20 pm	Three messages sent to "OLYMPIC".
3.55 pm	One message received from "OLYMPIC".
4.00 pm	One message sent to "OLYMPIC".
4.50 pm	One message recvd from "OLYMPIC".
5.37 pm	Two mges sent to "OLYMPIC".
7.00 p.m.	Sent names of First and Second Class survivors to "OLYMPIC".

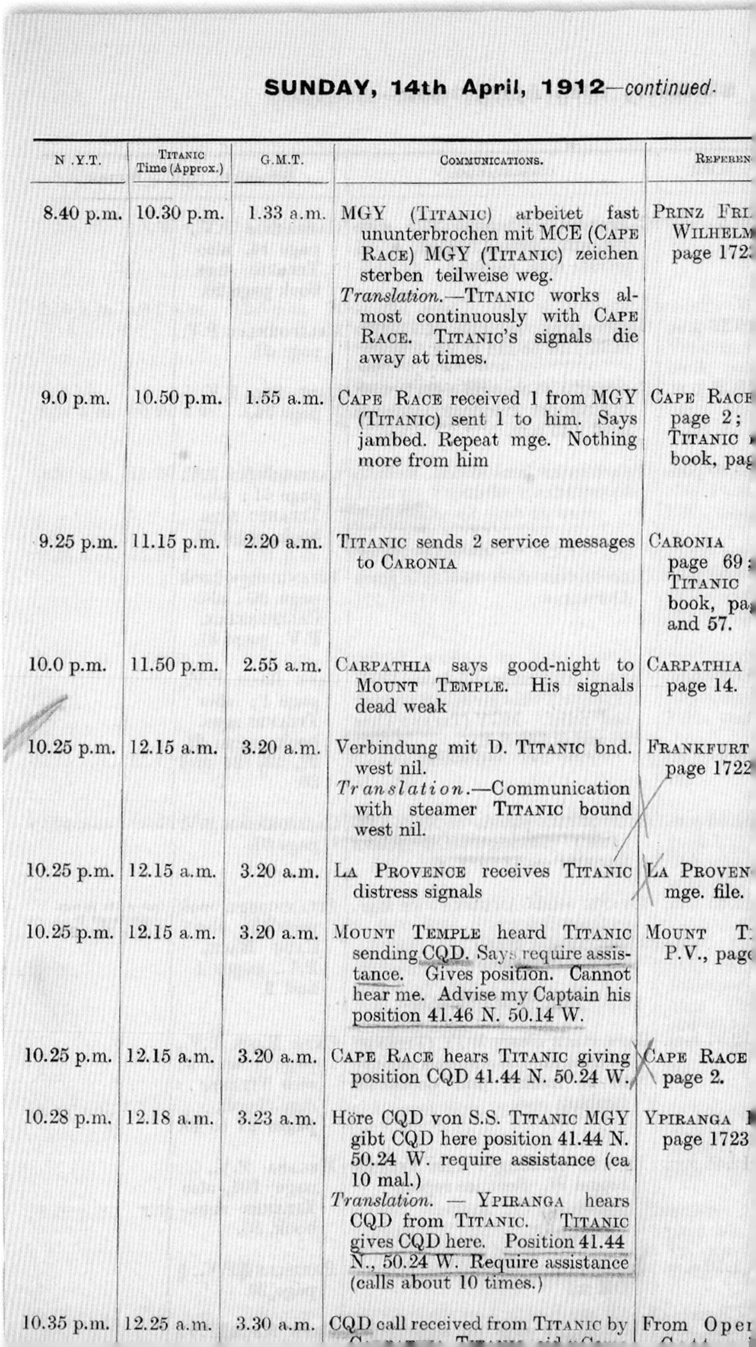

SUNDAY, 14th April, 1912—*continued*.

N.Y.T.	TITANIC Time (Approx.)	G.M.T.	COMMUNICATIONS.	REFEREN
8.40 p.m.	10.30 p.m.	1.33 a.m.	MGY (TITANIC) arbeitet fast ununterbrochen mit MCE (CAPE RACE) MGY (TITANIC) zeichen sterben teilweise weg. *Translation.*—TITANIC works almost continuously with CAPE RACE. TITANIC's signals die away at times.	PRINZ FRI WILHELM page 172
9.0 p.m.	10.50 p.m.	1.55 a.m.	CAPE RACE received 1 from MGY (TITANIC) sent 1 to him. Says jambed. Repeat mge. Nothing more from him	CAPE RACE page 2; TITANIC book, pag
9.25 p.m.	11.15 p.m.	2.20 a.m.	TITANIC sends 2 service messages to CARONIA	CARONIA page 69; TITANIC book, pa and 57.
10.0 p.m.	11.50 p.m.	2.55 a.m.	CARPATHIA says good-night to MOUNT TEMPLE. His signals dead weak	CARPATHIA page 14.
10.25 p.m.	12.15 a.m.	3.20 a.m.	Verbindung mit D. TITANIC bnd. west nil. *Translation.*—Communication with steamer TITANIC bound west nil.	FRANKFURT page 1722
10.25 p.m.	12.15 a.m.	3.20 a.m.	LA PROVENCE receives TITANIC distress signals	LA PROVEN mge. file.
10.25 p.m.	12.15 a.m.	3.20 a.m.	MOUNT TEMPLE heard TITANIC sending CQD. Says require assistance. Gives position. Cannot hear me. Advise my Captain his position 41.46 N. 50.14 W.	MOUNT T P.V., page
10.25 p.m.	12.15 a.m.	3.20 a.m.	CAPE RACE hears TITANIC giving position CQD 41.44 N. 50.24 W.	CAPE RACE page 2.
10.28 p.m.	12.18 a.m.	3.23 a.m.	Höre CQD von S.S. TITANIC MGY gibt CQD here position 41.44 N. 50.24 W. require assistance (ca 10 mal.) *Translation.* — YPIRANGA hears CQD from TITANIC. TITANIC gives CQD here. Position 41.44 N., 50.24 W. Require assistance (calls about 10 times.)	YPIRANGA page 1723
10.35 p.m.	12.25 a.m.	3.30 a.m.	CQD call received from TITANIC by	From Oper

Pages from the subsitute for the procès-verbaux of the *Titanic* and *Carpathia*, compiled for the British Inquiry into the sinking. (Oxford, Bodleian Library, MS. Marconi 258, fols. 182v–183r.)

SUNDAY, 14th April, 1912—*continued.*

N.Y.T.	TITANIC Time (Approx.)	G.M.T.	COMMUNICATIONS.	REFERENCE.	REMARKS.
10.35 p.m.	12.25 a.m.	3.30 a.m.	CAPE RACE hears MGY (TITANIC) give corrected position 41.46 N. 50.14 W. Calling him, no answer	CAPE RACE P.V., page 2.	
10.36 p.m.	12.26 a.m.	3.31 a.m.	MGY (TITANIC) sagt (says) CQD Here corrected position 41.46 N. 50.14 W. Require immediate assistance. We have collision with iceberg. Sinking. Can nothing hear for noise of steam. Position ca (sent about) 15 to 20 times.	YPIRANGA P.V., page 17232.	from Y 26 words
10.37 p.m.	12.27 a.m.	3.32 a.m.	TITANIC sends following "I require assistance immediately. Struck by iceberg in 41.46 N. 50.14 W." (Our distance from TITANIC 738 miles E. CARONIA.)	CARONIA P.V., page 69.	from Y 12 words
10.40 p.m.	12.30 a.m.	3.35 a.m.	TITANIC gibt mir position und sagt Please tell ur Captain to come to our help. We are on ice." *Translation.*—TITANIC gives his position to FRANKFURT and says "Tell your Captain to come to our help. We are on ice."	FRANKFURT P.V., page 17228.	from Y 18 words
10.40 p.m.	12.30 a.m.	3.35 a.m.	CARONIA sent CQ message to MBC BALTIC and CQD: "MGY (TITANIC) struck iceberg require immediate assistance"	CARONIA P.V., page 59.	Caronia to Baltic 9 words
10.40 p.m.		3.35 a.m.	MOUNT TEMPLE hears MGY (TITANIC) still calling CQD. Our Captain reverses ship. We are about 50 miles off	MOUNT TEMPLE P.V., page 32.	
10.46 p.m.	12.26 a.m.	3.41 a.m.	DKF (PRINZ FRIEDRICH WILHELM) ruft (calls) MGY (TITANIC) und gibt (and gives) position at 12.0 a.m. 39.47 N. 50.10 W. MGY (TITANIC) sagt (says) "Are u coming to our ?" DFT (FRANKFURT) sagt (says) "What is the matter with u ?" MGY (TITANIC): "We have collision with iceberg. Sinking. Please tell Captain to come." DFT (FRANKFURT) sagt (says), "OK, will tell ?"	YPIRANGA P.V., page 17232.	from Y 36 words
10.48 p.m.	12.38 a.m.	3.43 a.m.	MOUNT TEMPLE hears FRANKFURT give MGY (TITANIC) his position 39.47 N. 52.10 W.	MOUNT TEMPLE P.V., page 32.	
10.50 p.m.	12.40 a.m.	3.45 a.m.	OLYMPIC hears MGY (TITANIC) sig-	OLYMPIC P.V.,	

7.40 pm	Five mges recvd from "OLYMPIC".
8.25 pm	Sent three mges to "OLYMPIC" for re-transmission to New York via Cape Race.
8.30 pm	Sent two mges, to CUNARD NEW YORK & LIVERPOOL via "OLYMPIC" and CAPE RACE.
8.30 pm	"OLYMPIC" asks for names of Third Class Survivors, told him list not yet ready but will send them to you as soon as possible.
11.10 pm	Sent one mge to "CALIFORNIAN" for re-transmission to the "BALTIC".
	Tuesday 16 April 1912
10.25 am	One mge recvd from "CALIFORNIAN".
10.50 am	One mge sent to "CALIFORNIAN".
11.15 am	Trs with "MINNEWASKA", two mges recvd.
12.40 pm	One mge sent to "MINNEWASKA".
1.30 pm	Sent list of survivors names to "MINNEWASKA" for re-transmission to Sable Island "MINNEWASKA" says SABLE ISLAND not able to read me now.
5.30 pm	Three mges recvd from Sable Island.
	Wednesday 17 April 1912
5.00 am	One mge recvd from Sable Island.
5.25 am	Trs with "FRANCONIA" bound East.
7.40 am	22 mges sent to "FRANCONIA" for Sable Island.
9.15 am	One mge recvd from U.S.S. Chester.
10.0 am	One mge recvd from U.S.S. Chester.
10.15 am	One mge recvd from "KRONPRINZ WILHELM".
11.20 am	One mge recvd from U.S.S. Chester.
11.50 am	11 mges recvd from "FRANCONIA".
1.40 pm	15 mges sent to "FRANCONIA" for Sable Island.
2.40 pm	2 mges recvd from "BLUCHER".
3.10 pm	10 mges sent to Blucher for Cape Sable.

7.00 pm	One mge recvd from Siasconset.
	Thursday 18 April 1912
4.00 am	Five mges recvd from Siasconset.
6.35 am	25 mges sent to Siasconset.
8.30 am	29 mges sent to Siasconset.
Noon	In touch with Sagaponack.
12.20 pm	4 mges recvd from Sagaponack.
1.15 pm	16 mges sent to Sagaponack.
4.20 pm	23 mges sent to Sagaponack.
6.00 pm	In touch with Seagate.
7.10 pm	25 mges sent to Seagate.
7.45 pm	Two mges recvd from Seagate.
8.15 pm	Two mges recvd from Seagate.
9.40 pm	Two mges sent to Seagate.
9.45 pm	DOCKED AT NEW YORK.

S.S. *FRANKFURT*. PROCÈS-VERBAL

NYT	**Sunday 14 April 1912**[9]
9.35 pm	Exchange with DDC [Cincinnati].
10.05 pm	Call CQ – no response.
10.25 pm	Traffic with the Titanic [MGY] bnd west nil.
10.40 pm	Titanic gives me his position and says: please tell ur capt to come to our help we are on ice.
11.05 pm	MGY [Titanic] calling CQD and SOS. MKC [Olympic] and DDC [Cincinnati] answer.

11.10 pm	MGY [Titanic] continues calling CQD and SOS.
11.15 pm	Exchange with MGY [Titanic]. Tell him that we will take his course.
11.35 pm	MGY [Titanic] and MKC [Olympic] working together.
11.45 pm	Exchange with DYA [Ypiranga].
	Monday 15 April 1912
12.10 am	Traffic with the Birma [SBA] bnd east.
8.10 am	Arrived at the site of the disaster but MGY [Titanic] had already sunk 20 boats picked up by MAP [MPA: Carpathia] Searching for remains of ship.
9.00 am	Traffic with SBA [Birma].
9.45 am	Traffic with MAP [MPA: Carpathia].

S.S. *BALTIC*, PROCÈS-VERBAL

NYT	**Sunday 14 April 1912**[10]
7.50 pm	MGY [Titanic] finished MCE [Cape Race].
8.00 pm	Sigs MGY [Titanic].
8.10 pm	Sigs MPA [Carpathia].
8.25 pm	Sigs MCE [Cape Race].
9.40 pm	Sigs MCE [Cape Race].
9.55 pm	TRs Bohemian [MEL]: Am jambed [*sic*] badly. By on phones.
10.40 pm	Calling MGY [Titanic] No response.
11.10 pm	Jambing bad – but hear MGY [Titanic] very faint calling MKC [Olympic] – latter strong – freaky. Hear MPA [Carpathia] calling – he tells me MGY [Titanic] requires immediate assistance – gives pos as 41.46 N – 50.14 W – I advise Bridge + call MGY [Titanic] but unable to gain his attention – He appears to be tuned to MKC [Olympic] + cuts me out –
11.20 pm	3rd Officer in Cabin re my message to Commander.

11.35 pm MGY [Titanic] tells MKC [Olympic] "We are getting the women off in small boats" – MPA [Carpathia] tells MGY [Titanic] – Baltic coming to assistance – I don't appear to reach him.

11.45 pm MGY [Titanic] says "Engine room getting flooded".

Monday 15 April 1912

Message from Bridge gives our position 243 ~~234~~ (corrected by captain) miles East of MGY [Titanic]'s position –

12.10 am Sigs DDR [Amerika] + DKF [Prinz Friedrich Wilhelm] –

1.05 am MGN [Virginian] now calling MGY [Titanic] + MKC [Olympic].

1.10 am Sigs Virginian [MGN].

2.10 am DKF [Prinz Friedrich Wilhelm] calls CQ I tell him std by on phones.

2.15 am TRs Frankfurt [DFT].

3.05 am WQ [Eastport, Me.] asks DFT [Frankfurt] re CQD calls (He was jambing us at 11.5p).

5.05 am Sigs MPA [Carpathia] – Unable to work owing to persistent jambing by MWL [Californian] who is talking all time.

5.30 am MWL [Californian] persists in talking to SBA [Birma]. Such remarks as "Do you see a four masted salmon pink smoke stack steamer around" etc. Imposs for us to work MPA [Carpathia].

5.40 am MJL [Antillian] calls CQ – Told to std bi.

5.55 am Sigs MPA [Carpathia] but can do nothing for jambing by MWL [Californian] + SBA [Birma] who are carrying on long irrevelant [*sic*] conversations.

7.10 am In communication with MPA [Carpathia] exchange traffic re passengers + get instructions to proceed Liverpool. We turn round at 7.15a ~~about 7.20am~~ – We have come West 134 miles.

9.10 am Sigs DUS [United States].

9.20 am 2 from MGN [Virginian].

9.45 am Call MJL [Antillian] No response. (Brewer[a]).

10.15 am Sigs MGN [Virginian] + MJL [Antillian].

10.20 am	TRs Macedonia [MDT].
10.30 am	MWL [Californian] still monopolizing the air with his remarks – carrying on conversations with every station MPA [Carpathia] is trying to send me message, but communication out of the question owing to MWL [Californian].
11.00 am	Still imposs work MPA [Carpathia], owing MWL [Californian].
Noon	By on phones trying for MPA [Carpathia].

a Probably the junior wireless operator on S.S. *Baltic*.

R.M.S. *CARONIA*, PROCÈS-VERBAL

Ship's time	NYT	**Monday 15 April 1912**[11]
4.15 am	11.15 pm	MGY [Titanic] to MKC [Olympic]: Capt says get your boats ready what is your position?
4.20 am	11.20 pm	MBC [Baltic] – MPA [Carpathia]: pse tell MGY [Titanic] we are making toward her.
4.24 am	11.24 pm	MBC [Baltic] – MGY [Titanic]: We are making for you keep in touch with us.
4.35 am	11.35 pm	MGY [Titanic]'s sigs now inaudible Several stns coming in.
4.40 am	11.40 pm	Still nothing from MGY [Titanic].
4.45 am	11.45 pm	Hear MGY [Titanic] tho sigs unreadable.
4.52 am	11.52 pm	Hear MGY [Titanic] tho sigs unreadable still.
4.55 am	11.55 pm	Hear the DFT [Frankfurt] wkg to MGY [Titanic] – DFT [Frankfurt] according to position 172 miles from MGY [Titanic] at time first SOS sent out by the latter.
5.00 am	12.00 am	MCE [Cape Race] to Some Stn? You are much nearer to him than Olympic – He is already putting women off in boats and he says weather is clear and calm – MKC [Olympic] position 40.32 N 51.18 West, putting on all speed towards Titanic.

5.10 am	12.10 am	MSC [Siasconset] to MLP [La Provence] – CQD MSG from Titanic via CQ: position 41.46 N 50.14 W We require assistance struck iceberg Titanic. MLP [La Provence] to MSC [Siasconset]: OK rd [received]. MSC [Siasconset] to MLP [La Provence]: OK – yesterday April 13th we passed several icebergs in Lat. 41.20 N from 49 to 50 W.
5.11 am	12.11 am	MLP [La Provence] to MSC [Siasconset]: OK rd.
5.16 am	12.16 am	MBC [Baltic] calling MKC [Olympic].
5.34 am	12.34 am	MBC [Baltic] calling CQ.
5.39 am	12.39 am	Frm MGN [Virginian]: "Here is confirmation for what we have already said MSG 1 6 2 Cape Race to Capt Gambell Titanic struck iceberg require immediate assistance her position 41.64 N 50.14 W (sigs become faint) – sinking and –
5.45 am	12.45 am	MBC [Baltic] calling MKC [Olympic].
5.49 am	12.49 am	From Capt DFT [Frankfurt] MSG 1 16 Ypiranga position 2 am 38.12 49.37 Have stopped at site of disaster Do you know if this is near MGY [Titanic].[a]
6.05 am	1.05 am	MSC [Siasconset] calling MGN [Virginian]. MGN [Virginian] to MSC [Siasconset]: 240 east at 11pm. When did she strike.
6.16 am	1.16 am	"firing rockets – there – we are firing rockets", from unknown station "here look out for rockets".
6.37 am	1.37 am	DKF [Prinz Friedrich Wilhelm] calling.
6.32 am	1.32 am	MSG to Smith Titanic fm Captain Baltic via MGN [Virginian]: Baltic coming we are 234 miles east Commander Send that via MKC [Olympic] on to MGY [Titanic] soon as possible.
7.00 am	2.00 am	Stations working in distance but unreadable.
7.10 am	2.10 am	CQ from DKF [Prinz Friedrich Wilhelm].
7.11 am	2.11 am	MLP [La Provence] calls DKF [Prinz Friedrich Wilhelm]. MLP [La Provence] de [from] MLC [Celtic] nil. MLP [La Provence] to MLC [Celtic]: "At this – I have heard the MBC [Baltic] sending this mge to MGN [Virginian] – To Captain Smith Titanic. Baltic coming we are 234

		miles east – Nobody has heard MGY [Titanic] since about two hours. OK tis MLC [Celtic]. OK DKF [Prinz Friedrich Wilhelm] is about 250 miles astern of us now he was calling CQ CQ few minutes ago but I could not get him Is possible he has not received the CQD message If you can get him please retransmit it on MLP [La Provence]. OK tis Will try and get DKF [Prinz Friedrich Wilhelm] MLC [Celtic]. OK last night I heard MCE [Cape Race] calling you with Ss MLP [La Provence]. OK will look out MLC [Celtic].
7.20 am	2.20 am	MLC [Celtic] calling DKF [Prinz Friedrich Wilhelm].
7.52 am	2.52 am	MLP [La Provence] de [from] MLC [Celtic]. MLC [Celtic] de MLP [La Provence] K.
		MLC [Celtic] asks MLP [La Provence] if he can get MSG to Titanic. MLP [La Provence] replies no Refers MLC [Celtic] to DKF [Prinz Friedrich Wilhelm] and MRA [Caronia].
8.20 am	3.20 am	Rcd MSG from MLC [Celtic].
8.30 am	3.30 am	Called MGY [Titanic] no reply.
8.35 am	3.35 am	Called MGY [Titanic] no reply.
8.45 am	3.45 am	Sent message via CQ to MGY [Titanic].
10.05 am	5.05 am	Rcd 1 from MLC [Celtic].
10.35 am	5.35 am	MLC [Celtic] desires message to MGY [Titanic] cancelled.
…	…	…
		Tuesday 16 April 1912
12.25 am	7.25 pm	Recd 615 wds + 1 private MGE fm ZZ [Poldhu, England] Sigs good Distance 720 miles.
1.10 am	8.10 pm	Sig MLS [La Savoie] nil.
1.20 am	8.20 pm	Spoke MLC [Celtic] asked him if anything further re MGY [Titanic] MLC [Celtic] says MGY [Titanic] went down only 800 persons saved as reported by MKC [Olympic].
1.45 am	8.45 pm	MRA [Caronia] to MLC [Celtic] via DKF [Prinz Friedrich Wilhelm]: Capts Compts Can you give us some information re MGY [Titanic].

1.55 am	8.55 pm	Reply that MGY [Titanic] struck iceberg in 41–46N 50–14W Sank 800 saved.
1.57 am	8.57 pm	Asked for nr Casualties & who went to assistance (Capt anxious also wants to reply to Cunard as requested by them.
2.20 am	9.20 pm	Sigs MGA [Mauretania] nil.
2.35 am	9.35 pm	MBC [Baltic] supplies the following at Capts request: 1800 Casualties 800 Saved fm MGY [Titanic] – MPA [Carpathia] & myself went to assistance Carpathia there first –

a This message unclear; translated from the original German.

R.M.S *CELTIC*, PROCÈS-VERBAL

Ship's time	NYT	**Monday 15 April 1912**[12]
3.55 am	10.55 pm	Hear Titanic [MGY] say "I require immediate assistance – position 41.46 N 50.14 W" MRA [Caronia] confirms this – Reported to Commander immediately (798 miles).
4.08 am	11.08 pm	Rec'd CQD mge from Titanic [MGY] via MRA [Caronia].
4.32 am	11.32 pm	MPA [Carpathia] tells MGY [Titanic] "MBC [Baltic] is making towards you."
5.21 am	12.21 am	TR La Provence [MLP] Clear MSG from.
7.23 am	2.23 am	MLP [La Provence] says "Nobody has heard MGY [Titanic] for about two hours – MLP [La Provence] asks me to send MSG to DKF [Prinz Friedrich Wilhelm] + also says MCE [Cape Race] has "S" for me – unable get DKF [Prinz Friedrich Wilhelm].
8.18 am	3.18 am	Sent MSG to MLP [La Provence].
8.30 am	3.30 am	Sent MSG to MRA [Caronia] for MGY [Titanic].
10.25 am	5.25 am	MRA [Caronia] says cannot get message through to MGY [Titanic] by any means – I told him to cancel it.

11.36 am	6.36 am	TR Prinz Fred^k W^m [DKF: Prinz Friedrich Wilhelm] East Sent CQD.
12.30 pm	7.30 am	No reply to calls.
1.20 pm	8.20 am	No reply to calls.
1.25 pm	8.25 am	Clear 1 from DKF [Prinz Friedrich Wilhelm].
2.23 pm	9.23 am	Clear 1 to MRA [Caronia].
2.30 pm	9.30 am	Sigs DKF [Prinz Friedrich Wilhelm] Nil.
2.40 pm	9.40 am	TR Corinthian [MKN] East Nil.
2.40 pm	9.40 am	TR Amerika [DDR] East Nil.

S.S. *ASIAN*, PROCÈS-VERBAL

NYT	**Sunday 14 April 1912**[13]
11.12 pm	Titanic [MGY] clg me and sent pos Want immediate asstce. OK answered him promptly Recd the pos as lat 51 46 long 50.14 Informed Capt Instructs me to get it repeated.
11.15 pm	Sable Island [MSD] clg Olympic [MKC] No ans.
11.17 pm	Received MGY [Titanic] pos lat 41 46 long 50.14 (corrected) Informed Capt.
11.29 pm	Olympic [MKC] clg Titanic [MGY] Heard Titanic [MGY] faintly.
11.30 pm	Sable Is [MSD] clg Olympic [MKC] No ans.
11.35 pm	Olympic [MKC] sends MSG to Titanic [MGY].
11.40 pm	Called Olympic [MKC] No ans.
11.43 pm	Cld Sable Island [MSD] No ans.
11.58 pm	Titanic [MGY] cld SOS No-one ans.
	Monday 15 April 1912
12.03 am	Called Titanic [MGY] No ans.

S.S. *ASIAN*, WIRELESS OPERATOR'S REPORT

The Marconi International Marine Communication Co. Ltd

[For Head Office] Asian Station
April 30th 1912[14]

Sirs,

I beg to furnish you a detailed report of my communications with the White Star Liner Titanic April 14th 1912.

It will be noticed in my Returns for voyage endings April 30/1912 that I had occasion to be in communication with the Titanic at 11.12 pm 75th meridian time April 14/1912.

I first heard the Titanic at 7.40 pm April 14/1912 working with the shore station Cape Race. This communication continued until 7.47 pm. I could not hear Cape Race, not being inside his wireless radius. The condition of the Titanic's signals were very faint, and if he had worked other than a disc-spark it would have been very difficult to hear him at all. I could not tell you the nature of the traffic.

At 8.46 pm April 14th/1912 the Titanic and Cape Race were again in communication the condition of signals being similar to the first occasion.

The Titanic and Cape Race were apparently in communication from 9.14 pm to 9.33 pm April 14/1912 condition of signals still remained the same.

At 9.34 pm April 14th/1912 the Titanic commenced to call Cape Race but there were several ships evidently with traffic began to call and so jumbled up the signals.

At 10.30 pm April 14th 1912 the Titanic again called Cape Race but to my knowledge there was no response as I did not hear the Titanic again till he called me at 11.12 pm.

I called Sable Island four times in ten minutes 11.0 pm to 11.10 pm April 14th/1912 but got no response.

At 11.12 pm April 14th/1912 the Titanic called me and flashed the following.

"Want immediate assistance position 41.6 N 50.14 W"

I received the latitude as 51.46 N owing to the faint signals.

The captain was instantly informed and tolde [*sic*] me to get the latitude repeated. I got the Titanics correct position at 11.17 pm and promptly informed the captain.

The captain thought it not advisable to turn back for the Titanic. The principle [*sic*] reasons that influenced the captain in coming to this decision were in the first place, the Asian's distance, namely 300 miles which would take the Asian about thirty hours to steam, and would also necessitate casting of [*sic*] the vessel she had in tow. Secondly, after finding out that the Olympic was acquainted with the Titanics distress and also after obtaining the Olympics position the captain at once saw that the Olympic would reach the Titanics position some hours before the Asian. Thirdly the captains knowledge of there being several ships in the neighbourhood including the following :–

Carpathia (Cunard Line) Virginian (Allan Line) Baltic (White Star Line) Antillian (Leyland Line) Iroquois (standard Oil 6°) La Bretagne (Cie Gen–Transatlantique) Hellig Olav (Scandinavian-American Line) Marengo (Wilson Line) and Frankfort (Norddeutscher Lloyd Line).

The captain was also influenced by the fact that the Olympic carried a more powerful wireless apparatus which would enable her to spread the news in the direction of the Titanics position much quicker than the Asian could. The captain also knew that the Titanic was in communication with Cape Race at 9.33 pm April 14th/1912.

The Olympic called the Titanic at 11.29 pm and I heard the Titanic faintly reply. At 11.35 pm April 14th/1912 I heard the Olympic transmit the following message apparently to the Titanic.

"4.30am G.M.T. 40.32 N 61.18 W are you steering southerly to meet us."

I did not hear the Titanic acknowledge receipt of the message.

I last heard the Titanic calling S.O.S. and C.Q.D. at 11.58 pm April 14th 1912 his signals being very faint.

I continued duty till 12.50 am April 15th 1912 and endeavoured to spread the news of the Titanics distress by calling CQ Sable Island and Olympic but got no response from anyone.

At 5.32 am April 15/1912 I was in communication with the Olympic and he informed me that he last heard the Titanic calling S.O.S. and C.Q.D. at 11.55 pm April 14th 1912. I also informed him of my endeavours to get a message though to Sable Island regarding the Titanic which I finally did at 7.20 am April 15th/1912.

Yours faithfully

James Q Howard

S.S. *YPIRANGA*, PROCÈS-VERBAL

NYT	**Sunday 14 April 1912**[15]
10.28 pm	Hear CQD from S.S. Titanic [MGY]. MGY [Titanic] gives: CQD here position 41.44 N 50.24 W require assistance (about 10 times).
10.34 pm	MGY [Titanic] and MKL [Asian].
10.36 pm	MGY [Titanic] says: CQD here corrected position 41.46 N 50.14 W require immediate assistance, we have collision with iceberg, sinking can nothing hear for noise of steam (Position c. 15–20 times.).
10.44 pm	MGY [Titanic] gives CQD and position.
10.46 pm	DFT [Frankfurt] calls MGY [Titanic] and gives position at 12 am 39.47 N 50.10 W. MGY [Titanic] says: are u coming to our? DFT [Frankfurt] says: What is the matter with u? MGY [Titanic] says: We have collision with iceberg, sinking, pse tell captain to come. DFT [Frankfurt] says: OK will tell – ?
10.52 pm	MKC [Olympic] and MGN [Virginian].
11.02 pm	MGY [Titanic] gives: SOS – CQD and position.
11.05 pm	DDC [Cincinnati] calls MGY [Titanic] and gives position 37.36 N 54.44 W.
11.14 pm	MKC [Olympic] calls MGY [Titanic].
11.15 pm	DFT [Frankfurt] calls MGY [Titanic].
11.18 pm	MKC [Olympic] working MGY [Titanic].
11.26 pm	DFT [Frankfurt] working MGY [Titanic]. DFT [Frankfurt] says: our captain will go for ur course. MGY [Titanic] says: ok tks tks.
11.28 pm	MGY [Titanic] calls MBC [Baltic].
11.30 pm	MKC [Olympic] calls MGY [Titanic].
11.34 pm	MKC [Olympic] working MGY [Titanic]. MKC [Olympic] gives position – 4.24 am GMT: 40.32 N 61.18 W.
11.37 pm	MGY [Titanic] says: – we are putting the women off in the boats.
11.38 pm	Call MGY [Titanic] and give position 38N 50W.
11.40 pm	MGY [Titanic] says: We are putting passengers off in small boats.

11.42 pm	MGY [Titanic] and MKC [Olympic]; tells him the same.
11.44 pm	MBC [Baltic] calls MGY [Titanic].
11.47 pm	MBC [Baltic] says: We are rushing to u.
11.49 pm	DFT [Frankfurt] calls MGY [Titanic].
11.50 pm	MGY [Titanic] says something including "flooded". Cannot understand anything else, [the signal] is too broken.
11.58 pm	DFT [Frankfurt] calls MGY [Titanic].
12.00 am	[He replies]: std bi – std bi – std bi – go away.
	Monday 15 April 1912
2.00 am	Have heard no more from MGY [Titanic] since 11.50 pm.
3.24 am	SBA [Birma] says: we are 30 mls SW of MGY [Titanic].

S.S. *PARISIAN*, PROCÈS-VERBAL

NYT	**Sunday 14 April 1912**[16]
8.15 pm	4 TRs Titanic OK Nil – Asked MGY [Titanic] if he could take mge for MCE [Cape Race]- OK will try if can hear you.
8.30 pm	Sent MSG MGY [Titanic] OK
9.05 pm	MGY [Titanic] sending traffic to MCE [Cape Race] – Captain Stains informs me, Expect sight Oil Tank at daybreak and wish to call you as soon as sight her Better turn in.
	Monday 15 April 1912
5.40 am	Hear weak sigs of MPA [Carpathia] or some station saying Titanic struck iceberg, Carpathia has passengers from lifeboats.
6.05 am	4 TRs Virginian OK nil –Informed Captain Stains what I heard passing between ships regarding Titanic, and he decided not to return as MPA [Carpathia] was there and Californian [MWL] was 50 miles astern of us, but requested me to stand by in case required.

6.55 am 4 TRs Iroquois [MEI] & Navahoe [MEN] OK MEI [Iroquois] informs us Deutschland picked up by Asian.

9.15 am 4 TRs Olympic [MKC] OK Make enquiries re MGY [Titanic] I told him what I overheard from SBA [Birma] Not MPA [Carpathia].

10.24 am 4 MSGs MKC [Olympic] OK.

S.S. *MINNEWASKA*, OPERATOR'S REPORT

S.S. Minnewaska.
Tuesday April 16th 1912.[17]

Messrs.
The Marconi International Marine Communication Co. Ltd.
London.

Gentlemen,

I beg to report as follows on the exchange of traffic between S.S. Minnewaska and S.S. Carpathia with extracts from P.V. pages 146–7. Copies of MSGs enclosed.

...

6.50 am. Ex Trs Sable Island. Dist. 205 mls. S. Rcd MSG. (No. 1/1)

8.35 am. Clear MSG to MSD [Sable Island]. (No. 1/1)

(No reply to MSD [Sable Island] calls until 12.50 am.)

11.15 am. Ex Trs Carpathia bnd West and send MSG. (No. 2/1)

11.25 am. Sigs MPA [Carpathia]. (He offered MSG containing list of names which Capt. Gates desired to forward to N.Yk. but I told him to Std. Bi. while I tried to get MSD [Sable Island]. I failed so informed Capt.)

11.35 am. Clear MSG to MPA [Carpathia]. (No. 3/2)

12.40 pm. Sigs MPA [Carpathia] and "bi mo".

12.50 pm. Sigs MSD [Sable Island]. (Said he had stood bi for me all morning but had not heard my calls.)

1.00 pm.	Clear MSG from MPA [Carpathia]. (No.1/2) Sigs MSD [Sable Island] nil. (MSD [Sable Island] asked me to tell MPA [Carpathia] that he had 25 (?) mges for him and to send V's. MPA [Carpathia] refused, saying that he had too much work on hand.)
1.15 pm.	Sigs MPA [Carpathia]. (He again refused to have anything to do with MSD [Sable Island].)
1.30 pm.	MPA [Carpathia] commences list of names without any preamble. (I failed to make him understand that MSD [Sable Island] would probably be unable to read me by the time he finished. It was difficult to read him as his sigs were very loud and broken. The operator was obviously tired out and his sending very erratic. Eventually I made him stop and called MSD [Sable Island].)
2.20 pm.	Sigs MSD [Sable Island]. (MSD [Sable Island] said he was unable to read me and to tell MPA [Carpathia] to call him. I did so but got no reply.)
3.40 pm.	MWL [Californian] calling MPA [Carpathia] for MSD [Sable Island]. No reply.
4.20 pm.	Sigs MWL [Californian]. (He asked me to call MPA [Carpathia]. Did so but still no reply.)
5.10 pm.	MPA [Carpathia] wkg MSD [Sable Island]. (MPA [Carpathia] was getting out of range and I got no further reply to my repeated calls.)
	...

I also beg to draw your attention to P.V. 147, dealing with the S.S. Californian's traffic, and to P.V. 148, 149, 150 and 151, dealing with the S.S. Mauretania's traffic and the unavoidable delay in forwarding the same.

Yours faithfully,

[signed] A.S. Rawlings. (O.C.)

No. 1/1 MINNEWASKA 16 APR 1912
Office of Origin: New York To: Commander Minnewaska

We are particularly anxious to ascertain from Carpathia complete list of names of survivors Titanic aboard Carpathia Do your utmost get in touch with her and if so spare no trouble to get us all names and wire them unless you find Carpathia has forwarded them to us Answer
Franklin

No. 1/1 MINNEWASKA 16 APR 1912
Office of Origin: Minnewaska To: Ismay New York

Wire received Trying to get Carpathia So far unable
Gates

No. 2/1 MINNEWASKA 16 APR 1912
Office of Origin: Minnewaska To: Captain Carpathia

Am anxious to forward to N.Y. names of passengers from Titanic on board your ship If you have not already done so please let me have full list Compliments
Gates

No. 3/2 MINNEWASKA 16 APR 1912
Office of Origin: Minnewaska To: Captain Carpathia

Am out of touch with shore stations Will you please forward list to Ismay New York as soon as possible
Gates

No. 1/2 MINNEWASKA 16 APR 1912
Office of Origin: Carpathia To: Capt Minnewaska

Only too anxious to get all names to shore Doing all possible Mr Ismay is on board Bergs & pack ice down to 41.16 N and as far east as 49.30W
Rostron

SABLE ISLAND, PROCÈS-VERBAL

Local time	NYT	**Monday 15 April 1912**[18]
10.30 am–2.20 pm	9.30 am–1.20 pm	2 fm 1 to HX [Camperdown, Halifax] Commence list of survivors fm MKC [Olympic] Lose him Clear 500 words of it, then as MKC [Olympic] unable read us gives it to MCE [Cape Race].
…	…	…
		Tuesday 16 April 1912
…	…	…
1.10 pm	12.10 pm	MPA [Carpathia] clg MMW [Minnewaska] Call MPA [Carpathia] NR [no response].
1.12 pm	12.12 pm	MPA [Carpathia] sendg tfc MMW [Minnewaska].
1.50 pm	12.50 pm	2 fm HX [Camperdown] tis.
2.00 pm	1.00 pm	1 fm HX [Camperdown].
2.30 pm	1.30 pm	SB [Cape Sable] & HX [Camperdown] wkg 2 to MZN [Parisian].
2.31 pm	1.31 pm	1 fm MZN [Parisian].
2.40 pm	1.40 pm	1 fm HX [Camperdown] & 1 to tis.
2.50 pm	1.50 pm	Calg MPA [Carpathia] & MMW [Minnewaska] NR [no response].
3.20 pm	2.20 pm	Trs MWL [Californian] 170 SE 1 msg. MMW [Minnewaska] calls but now very distant & weak.
3.58 pm	2.58 pm	5 to MWL [Californian] for MPA [Carpathia] Stdg bi for MSB [Cape Sable].
4.16 pm	3.16 pm	1 to MZN [Parisian] & 2 fm tis.
4.25 pm	3.25 pm	HX [Camperdown] & MZN [Parisian] wkg tis.
4.45 pm	3.45 pm	HX [Camperdown] & MZN [Parisian] wkg.
4.50 pm	3.50 pm	1 to HX [Camperdown].
5.48 pm	4.48 pm	3 to MPA [Carpathia] Sent Nr 4 but NR Call MPA [Carpathia] NR.
5.55 pm	4.55 pm	2 fm MWL [Californian].

6.17 pm– 9.00 pm	5.17 pm– 8.00 pm	4 to 3 fm HX [Camperdown]. 73 fm HX [Camperdown] & bi while tell him of difficulties before and call MPA [Carpathia] No reply Other boats also clg him with rush msgs Jmg & X's continuous.
9.10 pm	8.10 pm	MWL [Californian] answers but HX [Camperdown] jms.
9.12 pm	8.12 pm	MWL says MPA [Carpathia] not now on Says clg him long time.
9.20 pm	8.20 pm	Sending one to MWL [Californian] Has X's & HX [Camperdown] jmg wkg MKL [Asian].
9.30 pm	8.30 pm	One to MWL [Californian]. HX [Camperdown] still jmg wkg MZC [Megantic] Can do nothing while this continues Shut.
9.50 pm	8.50 pm	HX [Camperdown] now wkg ANM [Minia] Boat clg us.
10.00 pm	9.00 pm	Ditto HX [Camperdown] & ANM [Minia] monopolising line Boat clg us.
10.05 pm	9.05 pm	Now through.
10.07 pm	9.07 pm	One fm MWL [Californian] Says MPA [Carpathia] not bi and won't have him in mng [morning] and that he has turned, hardly worth sending hundred mges to him on such a small chance.
10.10 pm	9.10 pm	One to HX [Camperdown], tell him conditions.
10.20 pm	9.20 pm	2 fm HX [Camperdown].
10.36 pm	9.36 pm	1 to MWL [Californian].
11.20 pm	10.20 pm	9 fm HX [Camperdown] & MWL [Californian] calls – try read fm MWL [Californian] but hopeless by jmg by MKC [Olympic] freaks.
		Wednesday 17 April 1912
12.00 am	11.00 pm	One fm MWL [Californian] & two to HX [Camperdown].
12.06 am	11.06 pm	Two fm HX [Camperdown] & ans MEA [Franconia].
12.14 am	11.14 pm	Trs MVO [Oceania] nil 110 SE bnd W.
1.30 am	12.30 am	37 fm HX [Camperdown] Call MPA [Carpathia] & CQ at intervals No reply Absolutely no sigs.
1.47 am	12.47 am	NDG [U.S.S. Chester] clg MPA [Carpathia] for long stretch No reply.

1.55 am	12.55 am	Hear MEA [Franconia] Call him but he wks MSC [Siasconset].
2.00 am	1.00 am	Lightning – aerial grounded.
2.10 am	1.10 am	Storm clearing, nothing doing.
2.35 am	1.35 am	Hear MEA [Franconia] Call him with tfc for MPA [Carpathia] No reply.
3.15 am	2.15 am	Start long code fm HX [Camperdown] Aerial charges heavily – grounded.
3.40 am	2.40 am	Clearing Call HX [Camperdown].
4.30 am	3.30 am	20 fm HX [Camperdown] No sign of MPA [Carpathia] Hear MEA [Franconia] but not tuned for us.
4.40 am	3.40 am	I fm HX [Camperdown].
5.30 am	4.30 am	MPA [Carpathia] calls CQ Ans him, sigs not quite readable.
5.36 am	4.36 am	I to MPA [Carpathia] & he offers "S" Tell him we have over hundred for him. MEA [Franconia] calls MPA [Carpathia] & jmg.
5.45 am	4.45 am	Call MPA [Carpathia] NR. MEA [Franconia] calls says will help you out with your tfc.
5.47 am	4.47 am	MEA [Franconia] can't raise MPA [Carpathia] now.
5.55 am	4.55 am	MEA [Franconia] wkg MPA [Carpathia] Cant hear MPA [Carpathia] now.
6.18 am	5.18 am	4 fm MEA [Franconia] tis.
6.50 am	5.50 am	MEA [Franconia] wkg MPA [Carpathia]. 4 to HX [Camperdown] tis.
7.00 am	6.00 am	MEA [Franconia] still wkg MPA [Carpathia] rec'g.
7.50 am	6.50 am	NDG [Chester] jmg MEA [Franconia] contly.
8.00 am	7.00 am	One to MEA [Franconia]. NDG [Chester] jmg him.
8.45 am	7.45 am	SB [Cape Sable] wkg HX [Camperdown].
8.50 am	7.50 am	Clg tfc to MEA [Franconia] Belt on engine slips When come back find SB [Cape Sable] trying receive one fm MEA [Franconia].
9.15 am	8.15 am	I to HX [Camperdown].

9.30 am	8.30 am	Trying send tfc to MPA [Carpathia] via MEA [Franconia] who has great difficulty reading for X's Stop Commence take his at 9.30 as appearing only way get anything done. NDG [Chester] jmg MEA [Franconia] all time.
10.30 am	9.30 am	20 fm MEA [Franconia] He still has X's Same here but going OK.
10.50 am	9.50 am	Balance fm MEA [Franconia] and bi while he trys MPA [Carpathia] One to MEA [Franconia], ten sendings.
11.05 am	10.05 am	SB [Cape Sable] wkg SJ [St John, New Brunswick]. DKF [Prinz Friedrich Wilhelm] now calls with "S".
11.10 am	10.10 am	Nothing else doing.
11.11 am	10.11 am	MEA [Franconia] wkg MPA [Carpathia] with msg then offers rush tfc fm Arcon which we are instructed to give precedence.
11.28 am	10.28 am	MEA [Franconia] asks who Arcon is Tell him.
11.30 am	10.30 am	MEA [Franconia] clearing another Arcon mge to MPA [Carpathia]. MSB [Cape Sable] endeavouring to butt in with SG to MEA [Franconia] Jmg him and likely MPA [Carpathia].
11.35 am	10.35 am	MEA [Franconia] clrg next Arcon mge Wkg easier now.
11.37 am	10.37 am	Now clrg Ist rush mge to MPA [Carpathia].
11.40 am	10.40 am	MEA [Franconia] now clrg his own tfc to MPA [Carpathia] Has two MSB [Cape Sable] continues jmg Tell HX [Camperdown] so.
11.55 am	10.55 am	One fm HX [Camperdown] rush Says better let balance outward wait bi mo while HX [Camperdown] clrs msg fm SB [Cape Sable].
Noon	11.00 am	HX [Camperdown] answering SB [Cape Sable] who won't go ahead.
12.03 pm	11.03 am	Call MEA [Franconia] NR. 64 fm HX [Camperdown]. DKP [Kronprinz Wilhelm] clg MPA [Carpathia].
12.05 pm	11.05 am	MEA [Franconia] says bi mo.
12.30 pm	11.30 am	Six to HX [Camperdown] and bi as MEA [Franconia] agn on with MPA [Carpathia].
12.45 pm	11.45 am	MEA [Franconia] still recg MPA [Carpathia].

12.50 pm	11.50 am	HX [Camperdown] with SB [Cape Sable].
1.25 pm	12.25 pm	MEA [Franconia] wkg MPA [Carpathia].
1.40 pm	12.40 pm	MEA [Franconia] gives MPA [Carpathia] bi till he clrs sum fm MSD [Sable Island].
1.55 pm	12.55 pm	DDB [Blucher] Clg MEA [Franconia] & tells him std bi till he clrs msg to CC [Cape Cod] Call MEA [Franconia] he says cant wk now bi tis.
2.12 pm	1.12 pm	4 to MEA [Franconia] & he wants more.
2.15 pm	1.15 pm	Ask MEA [Franconia] if any tfc for HO [Head Office] yet in order give precedence but he has not yet though Arcons mges deld to MPA [Carpathia] some time ago Tells us give him ten for MPA [Carpathia].
3.00 pm	2.00 pm	11 to MEA [Franconia] for MPA [Carpathia] tis.
3.08 pm	2.08 pm	MBT [Elizabethville] clg with tfc. MEA [Franconia] jmg him wkg MPA [Carpathia].
3.10 pm	2.10 pm	Trs DKP [Kronprinz Wilhelm] 240 mls SW nil. SB [Cape Sable] jmg wkg DDB [Blucher] preventing us work DKP [Kronprinz Wilhelm] with MPA [Carpathia] tfc.
3.20 pm	2.20 pm	Call MEA [Franconia] for tfc as he doing nothing No reply.
3.25 pm	2.25 pm	SB [Cape Sable] still wkg & MEA [Franconia] copying fm MWL [Californian].
3.30 pm	2.30 pm	SB [Cape Sable] jmg by numerous rq's fm DDB [Blucher] & holding up everything Tell HX [Camperdown] and try raise DKP [Kronprinz Wilhelm] or MEA [Franconia] No reply.
3.35 pm	2.35 pm	HX [Camperdown] tells SB [Cape Sable] stand bi He continues wkg DDB [Blucher]. MEA [Franconia] after MPA [Carpathia] who does not seem to be bi.
3.40 pm	2.40 pm	Commence clear inward tfc to HX [Camperdown] as unable to raise any others.

3.45 pm	2.45 pm	3 to HX [Camperdown] & DKP [Kronprinz Wilhelm] comes on. SB [Cape Sable] still jmg so preventing commn with DKP [Kronprinz Wilhelm] who should be able to relay to MPA [Carpathia] if not jmd.
3.48 pm	2.48 pm	HX [Camperdown] tries get MSB [Cape Sable] stop but ignored.
3.50 pm	2.50 pm	MEA [Franconia] unable raise MPA [Carpathia] so tell MEA [Franconia] give us rest his tfc Will endeavour work DKP [Kronprinz Wilhelm] after. X's coming on.
4.35 pm	3.35 pm	15 fm MEA [Franconia] He now calls MPA [Carpathia].
4.40 pm	3.40 pm	2 fm DKP [Kronprinz Wilhelm] Says lost MPA [Carpathia] this mng early.
4.50 pm	3.50 pm	MBT [Elizabethville] clg but standing bi to give MEA [Franconia] chance clr to MPA [Carpathia].
5.30 pm	4.30 pm	Trs MGA [Mauretania] 16 mges.
5.40 pm	4.40 pm	16 fm MGA [Mauretania] Jmg & X's. SB [Cape Sable] continually jmg clg HX [Camperdown] in middle of long mge fm MGA [Mauretania]. HX [Camperdown] gives std bi and wks SB [Cape Sable] Standing bi for SB [Cape Sable].
6.50 pm	5.50 pm	MEA [Franconia] calls with "S" Give him bi mo. HX [Camperdown] with SB [Cape Sable]. X's at intvls.
7.15 pm	6.15 pm	SB still with HX [Camperdown]. MEA [Franconia] with DDB [Blucher].
7.55 pm	6.55 pm	Balance 18 fm two to MEA [Franconia] Has not got MPA [Carpathia].
8.40 pm	7.40 pm	Long mge fm MEA [Franconia].
10.27 pm	9.27 pm	29 to HX [Camperdown] 5 fm HX [Camperdown].
11.35 pm	10.35 pm	Nils MER [Royal Edward] Call MEA [Franconia] NR.
11.55 pm	10.55 pm	3 fm HX [Camperdown] Call MEA [Franconia] between He answers once then goes.
Midnite	11.00 pm	2 fm HX [Camperdown].

SIASCONSET, PROCÈS-VERBAL

NYT **Monday 15 April 1912**[19]

3.40 am WS [New London, Connecticut] asks if we know anything abt MGY [Titanic] striking iceberg off MCE [Cape Race] = Tell him no ntg here = tis.

3.56 am NAD [Boston, Mass., Navy yard] NAF [Newport, R.I., Torpedo station] = NAD [Boston, Mass.] had MGN [Virginian] who was making for MGY [Titanic] but NAD [Boston, Mass.] lost him = has ntg definite abt MGY [Titanic].

4.10 am NMN [U.S.S. North Carolina] NAC [Portsmouth, N.H., Navy yard].

5.00 am NAD [Boston, Mass.] NAF [Newport, R.I.].

6.00 am NMN [North Carolina] clg NAH [Brooklyn, N.Y., Navy yard] I wkg NAF [Newport, R.I.] I NAD [Boston, Mass.]. Tis MAM [Luisiana] – Time F nil 107 bnd NYK. Time OK nil.

6.10 am MAM [Luisiana] knows ntg of MGY [Titanic].

6.15 am Nils MSK [Sagaponack] tell him re. MGY [Titanic] on request.

7.00 am DDS [President Grant] clg MSK [Sagaponack] MCE [Cape Race].

. . . . . .

Tuesday 16 April 1912

10.51 am DHO [Hellig Olav] & MSK [Sagaponack] – someone saying Titanic sank cant make out who it is jmg bad fm Navy stns.

11.06 am MSB [Cape Sable] has not had MPA [Carpathia] yet. CC [Cape Cod] says heard MGY [Titanic] sank but not confirmed yet.

. . . . . .

Wednesday 17 April 1912

6.15 pm MPA [Carpathia] ans me & offers "S". BA [SBA: Birma] jmg and NAE [Cape Cod, Highland Light, Massachussetts, Naval station] & MK [NMK: New York, Naval station] Ask them all to pse std Bi but no good they refuse.

6.28 pm MMV [Mesaba] SI.

7.20 pm Little headway with MPA [Carpathia] Either he is jmd or we are – Navy ships & stns wont heed our request to std bi. NDG [U.S.S. Chester] & NAE [Cape Cod, Highland Light] being persistent.

7.27 pm Navy stn holding key down for 10 seconds at a time. NRZ [U.S.S. Salem] holding his key down 30 seconds at a time.

7.40 pm NRZ [Salem] & NDG [Chester] sending "Ofm" Ses will be clear in a min. OK pse hurry.

8.20 pm NRZ [Salem] & NDG [Chester] still at it Impossible to work here.

8.30 pm After being in one hour NRZ [Salem] tells NDG [Chester] "to go get an operator" We are waiting for them to clear with 95 msgs on hand MPA [Carpathia] reports very heavy tfc for MSC [Siasconset] but system paralized [*sic*] on account of Navy. MCC [Cape Cod] after MPA [Carpathia] No ans.

8.35 pm NRZ [Salem] now holding key down.

8.40 pm New oper on NDG [Chester] comes on and starts the "Ofm" over again This is certainly the limit for the U.S. Govt to send a boat out like this = NDG [Chester] sends each word twice.

9.00 pm They are still at it each word twice Now 100 mges waiting Mighty shame.

9.25 pm Completely tied up NRZ [Salem] – NDG [Chester] sending long list of names each word twice The U.S. Navy is a blot on the map, detrimental to Business interests and a menace to Humanity – Xs beginning to be troublesome.

9.50 pm Navy ships give "gn" [good night] to each other. Calling MPA [Carpathia] No ans.

10.05 pm Light rumble X's Calling MPA [Carpathia] No ans. CC [Cape Cod] CQ.

10.10 pm Ditto.

10.20 pm Can't raise MPA [Carpathia].

Thursday 18 April 1912

2.09 am Get MPA [Carpathia] but NRZ [Salem] comes in clg MPA [Carpathia] with msg.

2.20 am NRZ [Salem] wkg MPA [Carpathia] = MPA [Carpathia] just unreadable X's.

2.23 am NRZ [Salem] says "pse repeat Xs Xs".

2.25 am Clg MPA [Carpathia] No ans.

2.30 am	NRZ [Salem] calls asks if we read MPA [Carpathia] Says he "has impt msg for MPA [Carpathia] but MPA [Carpathia] won't take it". This seems to be the case with our own tfc – Over 110 on hand and yet we have been unable to raise MPA [Carpathia] since 9.50 pm & 2.09 am when after attracting his attention he disappears ignores our calls & offers of "msgs" "urgent msgs" etc – X's are quite bad but traffic can be worked off through it especially with the good signals we are offering to MPA [Carpathia] Looks like a case of laziness "don't care" or something of that sort.
2.48 am	Get MPA [Carpathia] again Offer the question "Is Astor Butts & Guggenheim on board" He replies "Astor is not aboard the others I dont know of". (HO [Head Office] advised accdgly).
[No time given]	Offer tfc MPA [Carpathia] says "There is no use my taking your tfc mine is two days old gn gn" Call MPA [Carpathia] & ask if he can read our signals "Yes can read you OK, but say OM [old man] I have not been to bed since Titanic went down I have over 300 msgs". Tell him make a try and less argument Offer "msg from Mr Marconi". He gives "g" clears it OK & gives "g".
3.50 am	Little headway with MPA [Carpathia], sent 5 RI. MPA [Carpathia] gives "By mo" Been gone 10 mins This man has evidently made up his mind to get clear of MSC [Siasconset] tfc (HO [Head Office] advised).
4.00 am	Calling MPA [Carpathia] to get another start No answer.
4.10 am	MPA [Carpathia] on again sending V's & fooling with his spark Unable to get any reply as to whether nr 6 is Rd [received] or not.
4.12 am	Clg MPA [Carpathia] Get no reply – X gtg little better If man on MPA [Carpathia] would take for an hour or two and then send some of his tfc something might be done = Reading MSC [Siasconset] at 80 miles or so is too hard for an opr of his calibre. I personally have read MSC [Siasconset] thru worse X's than this @ 300 miles (MSB [Cape Sable]).
4.15 am	MPA [Carpathia] will not ans us.
4.20 am	MAA [Carmania] clg MPA [Carpathia].
4.25 am	Nils MAA [Carmania] = clg MPA [Carpathia] No reply.
4.40 am	Clg MPA [Carpathia] evy few mins = no reply. MAA [Carmania] also clg MPA [Carpathia] but unable raise him.

4.46 am	MAA [Carmania] clg MPA [Carpathia].
4.50 am	Clg MPA [Carpathia] No ans Very slight X's No reason why traffic should not be handled at a fast clip now But what can you do with a man such as is on MPA [Carpathia].
4.55 am	Clg MPA [Carpathia] & giving "g" No ans.
4.57 am	MPA [Carpathia] tinkering with his spark.
5.01 am	MPA [Carpathia] makes a couple of V's but don't answer us.
5.10 am	Clg MPA [Carpathia] & gvg him g No reply – MAA [Carmania] after him.
5.18 am	Get MPA [Carpathia] started He got our #6 OK.
5.30 am	MPA [Carpathia] gone off again S6 R5.
5.43 am	MPA [Carpathia] bi postn [position].
6.00 am	MPA [Carpathia] sending very badly Is continually falling down and not bothering to repeat when he does so.
6.25 am	MPA [Carpathia] S6 R23.
6.35 am	MPA [Carpathia] bis 10 mins to C [see] Capt S6 R25. MAA [Carmania] jmg. PK [MPK: Princess Adelaide] & PJ [Point Judith, R.I.].
6.45 am	Tell MAA [Carmania] to std Bi & not jmg with clg unnecessarily.
7.00 am	MSK [Sagaponack] CQs Clg MPA [Carpathia] No reply.
7.07 am	MAA [Carmania] says has 4 for MPA [Carpathia] Will we take em? Tell him have over hundred here and cant get em off – tis Clg MPA [Carpathia] = Has wasted half an hour now – MAA [Carmania] unable raise him also.
7.12 am	Get JB [MJB: Oravia] nil Std him bi = MPA [Carpathia] on again Jmg by PS [DPS: Prinz Sigismund] & NH [Unclear which station intended here].
7.30 am	MPA [Carpathia] mo S7 R30. MPA [Carpathia] stopping after every mge for something.
7.32 am	PS [Prinz Sigismund] jmg.
7.52 am	MPA [Carpathia] mo S7 R41.
8.25 am	PS [Prinz Sigismund] jmg MPA [Carpathia] S7 R54.

8.30 am	JC Rlf G off [change of operator].
8.40 am	Held up again MPA [Carpathia] not there.
8.43 am	MAA [Carmania] nil.
9.26 am	Rlv JC [change of operator].
10.00 am	MJB [Oravia] 5 Bulls [Bulletins] S27 = MK [NMK: New York] jmg Unable make him std Bi.
10.25 am	MJB [Oravia]'s spark gone unreadable all broken up.
10.30 am	Ask JB [Oravia] fix his spk = Call MPA [Carpathia] S Rush No sign of MPA [Carpathia] Not stg bi to take his rush msgs.
10.40 am	Clg MPA [Carpathia] continuously with 2 Impt msgs sgd Franklin No answer This ship has not been in attendance since 8.32 am Disgusting work for a Marconi ship MJB [Oravia] spk [spark] still broken down.

CQD AND SOS MESSAGES

From around 10.15 p.m. when Captain Smith entered the wireless room and instructed the operators to call for assistance, until around 12.30 a.m. when the *Titanic*'s power finally gave out, Jack Phillips was continually sending distress calls. Many of these calls took the form of the standard Marconi distress signal, 'CQD', which was a call to all ships, usually sent with the position of the ship in distress. In addition to CQD, Phillips also sent 'SOS', which had been the official distress call since 1908 (under the terms agreed at the 1906 Berlin Convention on Radiotelegraphy) but which was infrequently used owing to the greater familiarity of most operators with the Marconi signal. Harold Bride, in his statement to the *New York Times*, written immediately after the *Carpathia* docked at New York, recalled joking with Phillips about the distress calls: '"Send SOS" I said, "It's the new call, and it may be your last chance to send it".'[20] Contrary to rumour, this was not the first sending of SOS, which had been first recorded in 1909 and used sporadically in the intervening three years. This was, however, one of the most famous uses of the signal and was no doubt responsible for its increasing use thereafter.

The messages on the following pages are only a fraction of the messages sent out by the *Titanic*; many would not have been preserved and not all were submitted for the Inquiries. These originate from just a few of the ships which received communications from the *Titanic* on the night of the sinking. In spite of this, they serve to illustrate the urgency and confusion of those hours. Much of the information contained here is repeated in the PVs, but these individual messages are nonetheless important for the insight they provide into how the captains of the passing ships received the news of the disaster.

right
Message from the *Olympic* to the *Titanic*, 14 April. (Oxford, Bodleian Library, MS. Marconi 263, fol. 55.)

below
Distress message from the *Titanic* picked up by the *Celtic*, 15 April. (Oxford, Bodleian Library, MS. Marconi 263, fol. 73.)

73

Form No. 3—100—6.2.12. Forwarding Charge Deld. Date 15 Apl

THE MARCONI INTERNATIONAL MARINE COMMUNICATION CO., LTD.

No. 1/2 Celtic OFFICE. 15 Apl 191

Prefix SOS Code Words

Office of Origin Titanic

Service Instructions: cqd de MGY

COPY

Office Rec'd from	Time Rec'd	By whom Rec'd
MLP	5/21 a m.	[illegible]
Messenger.	Time Sent Out. m.	By whom Sent.

To CQ

CQD require assistance position 41.46 N

50.14 W Struck iceberg Titanic

Received from La Provence

Form No. 1—100—18.5.11. Sent date April 14th

The Marconi International Marine Communication Company, Ltd.

WATERGATE HOUSE, YORK BUILDINGS, ADELPHI, LONDON, W.C.

No. 2612 Olympic OFFICE April 14th 1912

Prefix MSG Code Words 13

Office of Origin Olympic

Service Instructions:

COPY

CHARGES TO PAY.		
Marconi Charge ...		
Other Line Charge...		
Delivery Charge ...		
Total . . .		

Office sent to	Time sent	By whom sent
MGY	11.50 p.m.	JM

READ THE CONDITIONS PRINTED ON THE BACK OF THE FORM.

To: Commander Titanic

Am lighting up all possible boilers as fast as can Haddock

PLEASE ASK FOR OFFICIAL RECEIPT.

Code Addresses registered on 7 with Cable Companies are not available for messages through British Post Office Stations.

CONINGHAM BROS., Printers, etc., Limehouse, E.

[Unknown] 15 April 1912[21]

Office of Origin: Titanic To: CQD

Position 41.44N 50.14W Struck iceberg

[Unknown] 15 April 1912[22]

Office of Origin: Titanic To: CQD

Position 41.46N 50.14W Require assistance Struck iceberg

La Provence 15 April 1912[23]

Office of Origin: Titanic To: CQD

CQD Position 41.44N 50.24 Require assistance

CQD Correct position 41.46N 50.14 Require assistance Struck assistance

Provence 15 April 1912[24]

Office of Origin: Titanic To: CQD

CQD Position 41.44N 50.24 Require assistance

CQD Correct position 41.46N 50.14 Require assistance Struck assistance

La Provence | 15 April 1912[25]
Office of Origin: Titanic | To: CQD

Position 41.46N 50.14W Require assistance Struck iceberg

La Provence | 14 April 1912[26]
Office of Origin: Titanic | To: Capt

Position 41.46N 50.14W Require assistance Struck iceberg

Celtic | 15 April 1912[27]
Office of Origin: Titanic | To: CQ

CQD require assistance Position 41.46N 50.14W Struck iceberg Titanic
Received from La Provence.

Celtic | 14 April 1912 (11.30 am)[28]
Office of Origin: Titanic | To: CQ

CQD Require assistance Position 41.46N 50.14W Struck iceberg Titanic

Celtic | 15 April 1912[29]
Office of Origin: Titanic | To: Commander Celtic

CQD Titanic in 41.46N 50.14W

[Unknown] | 15 April 1912[30]
Office of Origin: Titanic | To: Celtic CQD

CQD Position 41.44N 50.24 Require assistance
CQD Correct position 41.46N 50.14 Require assistance Struck assistance

Baltic | 14 April 1912[31]
Office of Origin: Titanic | To: Baltic

41.46N 50.14W Sinking want immediate assistance (243 off pos)[32]

Baltic | 14 April 1912[33]
Office of Origin: Titanic | To: Baltic

Capt Smith reports in collision with iceberg Wants immediate assistance
Sinking rapidly Get boats ready
GWB [G.W. Balfour, Senior Wireless Operator, S.S. Baltic][34]

Baltic | 14 April 1912[35]
Office of Origin: Baltic | To: Capt Smith Titanic

Baltic coming We are 243 miles East
Commander

Olympic | 14 April 1912[36]
Office of Origin: Olympic | To: Commander Titanic

Am lighting up all possible boilers as fast as can
Haddock

Olympic | 14 April 1912[37]
Office of Origin: Olympic | To: Commander Titanic

4.24 am GMT 40.32N 61.18W Are you steering southerly to meet us
Haddock

Virginian | 15 April 1912[38]
Office of Origin: Virginian | To: Captain Californian

Titanic struck berg Wants assistance urgently Ship sinking Passengers in boats
His position Lat 41.46 Long 50.14
Gambell Commander

Californian | 15 April 1912[39]
Office of Origin: (MWL) Californian | To: (MGN) Virginian

Please give MSG on account MGY so as Capt can go off track down to MGY[40]

ICE MESSAGES

It was standard practice for all North Atlantic shipping to exchange news on the location of ice on the major routes, as a safety measure, and as a matter of courtesy. It has been asserted that the spring of 1912 was unusually cold and that the North Atlantic ice-fields therefore extended further south than expected in April. Captain Rostron of the *Carpathia* stated that the course taken by the *Titanic* was entirely reasonable and that the sinking was 'most exceptional'.[41] It is certainly the case that messages relating to ice were being frequently sent between ships in the days leading up to 14 April. A number of these messages, both those sent to the *Titanic* and those sent between other ships, relating to ice in the same region are included here.

In the US Inquiry into the sinking there was a great deal of interest in four ice messages sent to the *Titanic* on 14 April, by the *Mesaba*, *Amerika* and *Californian* (two messages), relating to ice in the immediate vicinity of the *Titanic*. Stanley Adames, operator of the *Mesaba*, submitted his copy of an ice warning sent to the *Titanic* with a note, made at the time, which appears to show the *Titanic* operator's reply. Both Harold Bride and the surviving officers of the *Titanic*, however, denied all knowledge of this and the *Amerika*'s message.[42] Wireless regulations stated that any message sent with the preface 'SG' (or 'MSG') must be passed to and acknowledged by the captain of the receiving ship. The messages from Captain Smith to the captains of *La Touraine* and *Baltic*, acknowledging their notifications of ice, show that this policy was adhered to by the *Titanic*. It is not clear whether the *Mesaba* and *Amerika* messages were in fact received by Jack Phillips but it appears that they never reached the captain or officers.

In the case of the first message from the *Californian*, Harold Bride testified to having at first ignored the call, being busy writing up the day's accounts, but later overheard the message as it was being transmitted to the *Baltic*. He stated that he then acknowledged receipt to the *Californian* and delivered the message to the bridge. None of the surviving officers recalled taking this message but all remembered being told of an ice field, which they believed to be some distance ahead.[43] The final significant message was sent by the *Californian* less than an hour before the *Titanic* struck the iceberg. It was to inform the *Titanic*, which was believed to be within 20 miles of the *Californian*, that the latter had stopped for the night and was surrounded by ice. The operator of the *Californian*, Cyril Evans, told the US Inquiry that he received the response 'Shut up, shut up, I am busy; I am working Cape Race.'[44] The *Californian*, being close to the *Titanic*, would have had much stronger signals than Cape Race, which Phillips was trying to receive from, and was drowning it out.

These messages became a significant source of controversy and there has been much speculation that Phillips (and presumably other Marconi operators) neglected such traffic in order to prioritise commercial messages. It would be fair to say that in 1912 the procedures for reporting and responding to ice warnings were not clearly established. The tragic fact is that it took a disaster on the scale of the *Titanic* to bring about an investigation into the unused potential of wireless communication in improving safety at sea.[45]

below
Message from commander of the *Baltic* to Captain Smith on the *Titanic*, 14 April. (Oxford, Bodleian Library, MS. Marconi 264, fol. 22.)

overleaf
Message from Captain Smith to the commander of the *Baltic*, 14 April. (Oxford, Bodleian Library, MS. Marconi 264, fol. 31.)

22

Form No. 1—100.—18.5.11.

Sent date April 14

The Marconi International Marine Communication Company, Ltd.

WATERGATE HOUSE, YORK BUILDINGS, ADELPHI, LONDON, W.C.

No. 21 Baltic OFFICE April 14th 1912

Prefix MSG Code LG Words 73

Office of Origin Baltic

Service Instructions: COPY.

CHARGES TO PAY.		
Marconi Charge ...		
Other Line Charge...		
Delivery Charge ...		
Total . . .		

Office sent to	Time sent	By whom sent
MGY	1 52 a m.	GWB

READ THE CONDITIONS PRINTED ON THE BACK OF THE FORM.

To: Capt Smith Titanic

Have had mod Var Winds and clear fine weather since leaving Greek Str Athinai reports passing Icebergs and large quantity of field Ice today in latitude 41-51 N Long 49-52 W Last night we spoke German oil tank steamer Deutschland Stettin to Philadelphia not under control short of coal Lat 40-42 N long 55-11 W Wishes to be reported

PLEASE ASK FOR OFFICIAL RECEIPT.

Code Addresses registered only with Cable Companies are not available for messages through British Post Office Stations.

CONINGHAM BROS., Printers, etc., Limehouse, E.

ICE WARNINGS SENT TO THE *TITANIC*

[Unknown]
Office of Origin: Touraine

12 April 1912[46]
To: Capt Titanic

My position 7 pm GMT Lat 49.28 Long 26.18 W. Dense fog since the night Crossed thick icefield Lat 44.58 Long 50 40 Paris [Parisian]. Saw another icefield and two icebergs Lat 45 20 Long 45 09 Paris [Parisian]. Saw a derelict Lat 40 56 Long 68 38 Paris [Parisian]. Please give me yr position Best regards and bon voyage

Coussin

La Touraine
Office of Origin: Titanic

12 April 1912[47]
To: Capt La Touraine

Thanks for your message and information My position 7pm GMT Lat 49.45 Long 23.38 W geenwick [Greenwich] Had fine weather Compts

Smith

Baltic
Office of Origin: Baltic

14 April 1912[48]
To: Capt Smith Titanic

Have had mod var [moderate variable] winds and clear fine weather since leaving Greek steamer Athinai reports passing icebergs and large quantity of field ice today in Lat 41–51 N Long 49–52 W Last night we spoke German oil tank steamer Deutschland Stettin to Philadelphia Not under control Short of coal Lat 40.42 N Long 55.11 W Wishes to be reported to New York and other steamers. Wish you and Titanic all success

Commander

Baltic
Office of Origin: Titanic

[14 April 1912][49]
To: Commander Baltic

Thanks for your message and good wishes Had fine weather since leaving

Smith

31

Form No. 3—100—6.2.12. *Forwarding Charge* ______ *Deld. Date* 14 April

THE MARCONI INTERNATIONAL MARINE COMMUNICATION CO., LTD.

No. 1/3 SS Baltic *OFFICE.* 14 April 191

Prefix MSG *Code* ______ *Words* 15

Office of Origin Titanic

Service Instructions:

COPY.

Office Rec'd from	Time Rec'd	By whom Rec'd
MGY	12.55 p. m.	GWB
Messenger.	Time Sent Out. m.	By whom Sent.

To Commander Baltic

thanks	for	your	message	and
good	wishes	had	fine	weather
since	leaving			
		Smith		

Caronia
Office of Origin: Noordam

14 April 1912[50]
To: Capt SS Titanic

Congratulations on new command Had moderate westerly winds Fair weather No fog Much ice reported in Lat 42.24 to 42.45 and Long 49.50 to 50.20 Compts
Krol

Mesaba
Office of Origin: Mesaba

14 April 1912[51]
To: Titanic [and] all Eastbound ships

Ice report In Lat 42N to 41.25N Long 49W to Long 50.30W saw much heavy pack ice and great number large icebergs also field ice Weather good clear

Reply Received thanks
Sent this to about 10 other ships as names in the PV
SHA[52]

OTHER ICE WARNINGS SENT AT THE TIME

Carmania — 11 April 1912[53]
Office of Origin: Carmania — To: Capt Caronia

4 am GMT Lat 41.45N Long 52.12W Had light to mod SW to NW winds since leaving Patches of fog from 48 to 51W Passed a large number of bergs growlers and extensive field ice in Lat 41.58N Long 50.20W Compts
Dow

Antillian — 13 April 1912[54]
Office of Origin: Antillian — To: Capt Wood Asian

Noon 40.04N 57.37W Run 280 miles No observation Ice bergs reported 41.50N 50.20W by President Grant bound West Regards
Japha

Corsican — 13 April 1912[55]
Office of Origin: Corsican — To: Captain Virginian

8 pm 44.3N 61.40W Passed heavy field ice and numerous bergs from 42.15N 49.48W to 41.25N 50.20W Had fog last of 49.30W Tunisian met heavy field ice Steamed 50 miles South Cleared ice in 43.02 49.30W but had foggy weather Regards
Cook

Caronia — 13 April 1912[56]
Office of Origin: Caronia — To: Captain Cincinnati

Westbound steamers report bergs growlers and field ice 42N from 49 to 51W Compts
Barr[57]

Amerika — 14 April 1912[58]
Office of Origin: Amerika via Titanic and Cape Race — To: Hydrographic Office Washington DC

Amerika passed two large icebergs in 41.27N 50.8W on the 14th of April
Knuth

Camperdown 14 April 1912[59]
Office of Origin: DDF [S.S. Pisa] To: F.S. Hydrographic Office Washington DC

In Lat 42.6 and Long 49.43 W met with extensive field ice and sighted seven bergs of considerable sizes on both sides of track
Captain

Caronia 14 April 1912[60]
Office of Origin: Caronia To: Captain Celtic

Westbound steamers report bergs growlers and field ice in 42N from 49W 51W April 12th Have had mod westerly winds and clear weather since leaving Regards
Barr

Celtic 14 April 1912[61]
Office of Origin: Celtic To: Capt Caronia

Thanks your message Had mod weather all the way Today Tunisian reports berg seen in 43N 40W Regards
Hambelton

Mesaba 14 April 1912[62]
Office of Origin: Mesaba To: Capt. Harris Parisian

Noon Lat. 42.02N Long. 49.25W Will let you know if I get her Now 5.30 Pm G.M.T. Lat 41.59W Long 50.02W Passing many large icebergs & field ice Compliments
Clarke

Mesaba 14 April 1912[63]
Office of Origin: Mesaba To: Capt Harris Parisian

Yes had to steer SW to clear end of ice which was in about Lat 41.35N Long 50.30W Now 11.30 pm G.M.T. Long 51.28W Weather nice and clear No ice in sight Regards
Clarke

Parisian 14 April 1912[64]
Office of Origin: Parisian To: Californian

Lat 41.55N 49.14W Passed three large bergs

[Unknown] 14 April 1912[65]
Office of Origin: La Bretagne To: Captains of Europe-bound ships

Met with ice-field and icebergs from 42N and 49 to 50W Compliments
Mace

Sent 14th April 1912 to Captain Marengo 8.08 am
Olympic 11.00
Campanello 12.00
Pennsylvania 14.08[66]

Californian 14 April 1912[67]
Office of Origin: Californian To: Capt Antillian

6.30 pm AST [Atlantic Standard Time] Lat 42.3N Long 49.9W Three large bergs five miles to southward of us Regards
Lord

Antillian 14 April 1912[68]
Office of Origin: Antillian To: Capt Lord SS Californian

7 pm AST [Atlantic Standard Time] 40.56N 50.22W Thanks for information Seen no ice Bon voyage
Japha

Minnewaska 14 April 1912[69]
Office of Origin: George Washington To: Capt S.S. Minnewaska

Foggy since Nantucket Field ice and bergs between 42.13N 49.40W and 41.52N 50.30W Compts
Polack

Celtic 15 April 1912[70]
Office of Origin: La Provence To: Captain Celtic

We passed several ice bergs in Lat 41.30N from 49 to 50W

Asian 15 April 1912[71]
Office of Origin: Asian To: Captain Olympic

13th April Iceberg reported in Lat 41.50N Long 50.20W Regards
Wood

OFFICIAL MESSAGES

Among the many people desperate for news when the first messages of the *Titanic*'s distress reached the shore were the companies with a direct interest in the welfare of their property and employees. White Star Line, owners of the *Titanic*, bombarded the *Olympic* (*Titanic*'s sister ship) and later the *Carpathia* with urgent messages from their office in New York. For some hours there was total confusion while the *Olympic* was unable to make direct contact with the *Carpathia*. Rumours abounded: that the *Titanic* was afloat and being towed into Halifax, that passengers were aboard a number of different ships, including the *Virginian* and *Parisian* as well as the *Carpathia*, and many more. Echoes of this can be seen in the desperate messages from Philip Franklin, vice president of the International Mercantile Marine Co. (owners of White Star Line and a number of other shipping lines), to Captain Haddock of the *Olympic*. When contact between the *Olympic* and the *Carpathia* was finally established, almost twelve hours after the *Titanic* had foundered, the first historic messages, with accurate news of the sinking, were transmitted. A number of these were preserved and the conversation can be

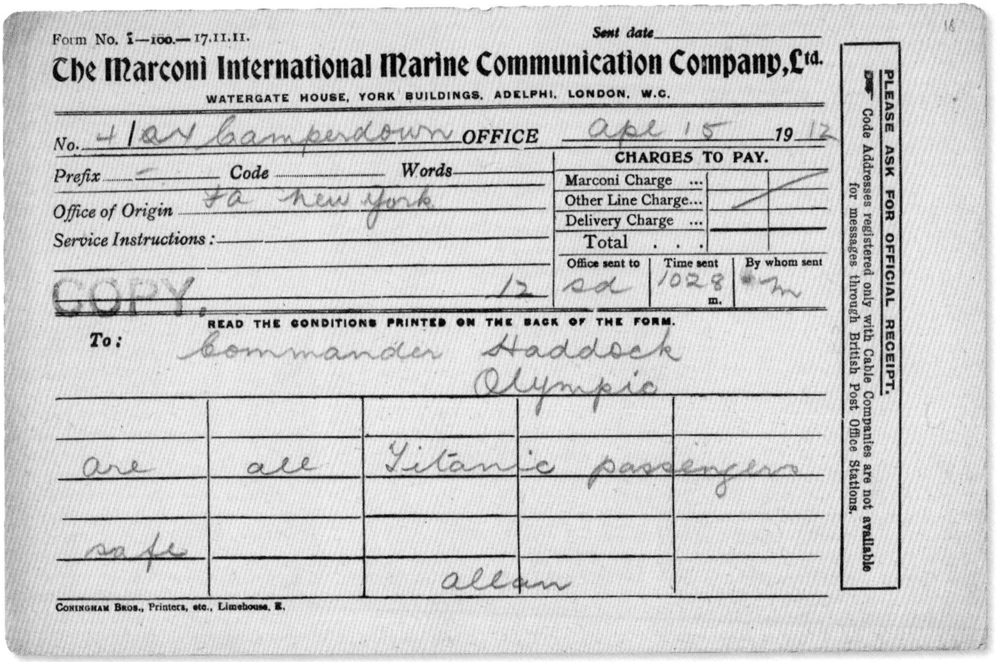

Form No. 1—100.—17.11.11. Sent date

The Marconi International Marine Communication Company, Ltd.

WATERGATE HOUSE, YORK BUILDINGS, ADELPHI, LONDON, W.C.

No. 4 / as Camperdown OFFICE apl 15 1912

Prefix — Code Words

Office of Origin Fa New York

Service Instructions:

COPY. 12

CHARGES TO PAY.		
Marconi Charge ...		
Other Line Charge...		
Delivery Charge ...		
Total . . .		

Office sent to	Time sent	By whom sent
sd	1028 m.	

READ THE CONDITIONS PRINTED ON THE BACK OF THE FORM.

To: Commander Haddock Olympic

are all Titanic passengers safe

allan

Coningham Bros., Printers, etc., Limehouse, E.

PLEASE ASK FOR OFFICIAL RECEIPT.

Code Addresses registered only with Cable Companies are not available for messages through British Post Office Stations.

Message from Allan, New York, to Captain Haddock on the *Olympic*, 15 April. (Oxford, Bodleian Library, MS. Marconi 266, fol. 18.)

followed (although not quite complete) through the messages printed here.

Once this information reached White Star Line and (via interception of messages by the American press) the general public, the *Carpathia* was inundated with requests for information. Messages recorded here are from White Star Line, the American and Canadian Marconi Companies, and Cunard Line (owners of the *Carpathia*). Many of these messages were substantially delayed by the volume of wireless traffic and the reluctance of the *Carpathia* to take the vast quantities of incoming messages until lists of survivors could be compiled and sent out. A substantial amount of the traffic being sent to the *Carpathia* related to a few high-profile passengers. The press, the shipping companies, the Marconi Company, national governments, and of course relatives, were desperate to know whether figures such as millionaire Col. John Jacob Astor IV, owner of Macy's department store Isidor Straus, and wealthy businessmen Benjamin Guggenheim, G.D. Widener, John Thayer and Charles Hays were on board. These enquiries were somewhat sporadically dealt with, leading to significant frustration. White Star Line resorted to sending telegrams signed 'Sumner' (the name of the Cunard Line agent in New York) in the hope that the *Carpathia* would at least respond to its own officers if not those of the *Titanic*.[72] There is no evidence that this achieved greater success, although two

Message from Captain Haddock, *Olympic*, to [White Star Line], 15 April. (Oxford, Bodleian Library, MS. Marconi 266, fol. 28.)

28

No. 1a.

MARCONI WIRELESS TELEGRAPH COMPANY OF CANADA, Limited

MONTREAL

No. 16 Campendown STATION Apl 15th 1912

Prefix S Code Words 21

Office of Origin MKC

Service Instructions:

Station Rec'd from	Time Rec'd	By whom Rec'd
SS	1126 a.m.	9n

Stations sent to	Time Sent	By whom sent
3ax	1131 a.m.	JC

To Ismay Newyork

Parisian reports Carpathia in attendance and picked up twenty boats of passengers and Baltic returning to give assistance position not given

Haddock

121

No. 1a.

MARCONI WIRELESS TELEGRAPH COMPANY OF CANADA, Limited

MONTREAL

No. Camperdown STATION Apr 17 1912

Prefix Code Words 18

Office of Origin MPA

Service Instructions: via MEC

via C.P.R.

Station Rec'd from	Time Rec'd	By whom Rec'd
SO.	7 44 am.	W.
Stations sent to	**Time Sent**	**By whom sent**
7	8 55 am.	MB.

To Islefrank New York

Deeply	regret	advise	you	titanic
sank	this	morning	after	collision
iceberg	resulting	serious	loss	life
further	particulars		later	
		Bruce	Ismay	

messages from the *Carpathia* to Cunard (both partially in code) are recorded here.

An intriguing set of messages are those between J. Bruce Ismay, managing director of White Star Line, on board the *Carpathia*, and Philip Franklin, his superior in New York. The first of these messages, the brief and poignant notification that the *Titanic* had sunk, became a subject of controversy on account of the fact that, although Ismay wrote it on the morning of 15 April, it was not received by Franklin until two days later. The cause of this is perhaps that it was treated as a passenger's message rather than an official message. Captain Rostron's instructions to Harold Cottam were: 'Official messages first. After they had gone, and the first press message, then the names of

Message from Bruce Ismay to the White Star Line, 17 April. (Oxford, Bodleian Library, MS. Marconi 266, fol. 121.)

the passengers. After the names of the passengers and crew had been sent, my orders were to send all private messages from the *Titanic*'s passengers first, in the order in which they were given to the purser; no preference to any message.'[73] No passengers' messages appear to have been sent before 17 April, so if Ismay's message was not prioritised the delay is understandable.

The subsequent conversation between Ismay and Franklin centred on the issue of how to return the surviving crew of the *Titanic* to England (at this time neither knew that a US Inquiry into the sinking would necessitate the officers and many of the crew remaining in New York for some days). Ismay, in these messages, insisted with considerable force that the SS *Cedric* be delayed in its departure from New York in order that the *Titanic*'s crew could be transferred to it immediately after the *Carpathia* docked. The explanation for this apparently surprising request was provided at the Inquiry by Second Officer C. H. Lightoller, who explained that the motivation for holding the *Cedric* was to ensure that the crew of the *Titanic* remained together. His concern was that, left to their own devices, the men would disperse, finding themselves positions on numerous other ships, and be untraceable in the event of an Inquiry being called back in the UK. Lightoller stated that, although Ismay was the official signatory on these telegrams (under his code-name 'Yamsi'), he was in fact in no condition to be making such decisions at the time, being extremely distressed with guilt at having survived.[74] Ismay made no reference to Lightoller's involvement or his own mental condition, but in his testimony

Message from Bruce Ismay to White Star Line, 17 April. (Oxford, Bodleian Library, MS. Marconi 266, fols 122–3.)

87

122

No. 1a.

MARCONI WIRELESS TELEGRAPH COMPANY OF CANADA, Limited

MONTREAL

No. 30 Camperdown *STATION* Apr 17 *19* 12

Prefix S *Code* *Words* 50

Office of Origin MPA via Mea

Service Instructions:

Via W.U.

Station Rec'd from	Time Rec'd	By whom Rec'd
SD	5 12 pm.	W
Stations sent to	**Time Sent**	**By whom sent**
33 Hy	6 pm.	P S M

To Islefrank New York

most desirable Titanic crew aboard Carpathia Should be returned home earliest moment posible Suggest you hold Cedrick sailing her daylight friday unless you see any reason kontrary propose returning in her myself please send outfit of Clothes including

No. 1a.

MAR

No.

Prefix

Office of Origin

Service Instructions:

Via W.U.

	Time Rec'd	
	m.	
Stations sent to	**Time Sent**	**By whom sent**
	m.	

To

shoes for me to Cedric have nothing of my own please reply

Yamsi

he seemed unable to remember with any certainty the order of the messages and the times they were sent.[75]

The final set of messages in this section relate to the recovery of the bodies of those who had died in the water. The first ship sent to *Titanic*'s location to search for bodies in the water was the cable ship *Mackay-Bennett*, which found so many that, in spite of instruction to the contrary, it was unable to preserve all of the bodies on board and was forced to bury some at sea, many unidentified. This controversial decision was made the more so by the fact that those buried at sea were almost all Third Class passengers and crew. The apparent reasoning behind this was that it was more important for wealthy passengers to be identified as there was likely to be greater dispute regarding the settlement of their estates. The second ship to arrive at the scene of the disaster was the cable ship *Minia*. Substantial correspondence relating to the despatch of the *Minia* is preserved and printed here. The *Minia* was first contacted about the possibility of returning to the *Titanic*'s position on 16 April, before it was known that the *Titanic* had sunk. Initial discussions, as shown, related to the ability of the *Minia* to assist a stranded ship. It was perhaps not until the *Minia* reached Halifax that it was known that the expedition would be simply for recovery of bodies. In total, 333 bodies were recovered, the vast majority of these by the *Mackay-Bennett*. Each body was numbered and, where possible, identified either by the contents of their pockets or by friends and relatives. Hundreds of bodies were never found.

OFFICIAL MESSAGES

Camperdown — 15 April 1912[76]
Office of Origin: New York — To: Capt Smith SS Titanic

Anxiously awaiting information and probably disposition passengers
Franklin
[White Star Line]

Carpathia — 15 April 1912[77]
Office of Origin: Carpathia — To: Cunard NYK

7.55 am NYK Lat 41.45N Long 50.20W Orfanello NYK unless otherwise ordered with about unpasieron Calmarers with Mr Ismay and bonplandie with so much ice about consider NYK best Large number icebergs and 20 miles field ice with bergs amongst
Rostron

Camperdown — 15 April 1912[78]
Office of Origin: New York — To: Capt Haddock SS Olympic

Endeavour communicate with Titanic and ascertain time and position Reply soon as possible to Ismay New York
F.W. Redway
[White Star Line]

Camperdown — 15 April 1912[79]
Office of Origin: MKC [Olympic] — To: Ismay New York

Since midnight when her position was 41 46N 50 14W have been unable to communicate We are now 310 miles from her nine am under full power Will inform you at once if hear anything
Commander

Camperdown — 15 April 1912[80]
Office of Origin: New York — To: Haddock Olympic via CQ

Do utmost ascertain immediately and advise us fully disposition Titanic passengers and where they will be landed
Franklin

Camperdown 15 April 1912[81]
Office of Origin: MKC [Olympic] To: Ismay New York

Parisian reports Carpathia in attendance and picked up twenty boats of passengers and Baltic returning to give assistance Position not given
Haddock

Camperdown 15 April 1912[82]
Office of Origin: New York To: Captain Haddock Olympic

Thanks your message. We have nothing from Titanic but rumored here that she proceeding slowly Halifax but we cannot confirm this We expect Virginian alongside Titanic Try communicate her
Franklin

Camperdown 15 April 1912[83]
Office of Origin: New York To: Commander SS Olympic

Keep us fully posted regarding Titanic
Franklin

Olympic 15 April 1912[84]
Office of Origin: Olympic To: Captain Carpathia

7.12 pm GMT Our position 41.17N 53.53W Shall I meet you and where
Haddock
Steering East true

Olympic 15 April 1912[85]
Office of Origin: Carpathia To: Olympic

We received distress signal call from the Titanic at eleven twenty and proceed right to spot mentioned On arrival at daybreak we saw ice 25 miles long apparently solid quantity of wreckage and number of boats full of lives We raised about six hundred and seventy souls Titanic has sunk She went down in two hours Captain and all engineers Our captain sent order that there was no need for Baltic to come any further so with that she returned on her course to Liverpool We have two or three officers aboard and the second Marconi operator who had been creeping his way through water at 30 degrees for several hours Mr Ismay is aboard

Olympic 15 April 1912[86]
Office of Origin: Olympic To: Captain Carpathia

Kindly inform me if there is the slightest hope in searching Titanic position at daybreak Agree with you on not meeting Will stand on present course until you have passed and will then haul more to southward Does parallel of 41.17N lead clear of the ice Have you communicated disaster to our people at New York or Liverpool or shall I do so and what particulars can you give me to send Sincere thanks for what you have done
Haddock

Carpathia 15 April 1912[87]
Office of Origin: Carpathia To: Olympic

South point pack ice in 41.16N Dont attempt to go north until 49.30W Many bergs large and small amongst pack Also for many miles to Eastward Fear absolutely no hope searching Titanics position Left Leyland SS Californian searching round All boats accounted for About 675 souls saved crew and passengers Latter nearly all women and children Titanic foundered about 2.20 am 5.47GMT in 41.16N 50.14W Not certain of having got through Please forward to White Star also to Cunard Liverpool and New York and that I am returning New York Consider this most advisable for many considerations
Rostron

Camperdown 16 April 1912[88]
Office of Origin: New York To: Commander SS Carpathia

What is your present position When do you expect reach New York Sumner Stop Anxiously awaiting names remaining additional survivors and crew
Franklin

Camperdown 16 April 1912[89]
Office of Origin: New York To: Commander Steamer Carpathia

Vitally important that we receive names balance survivors including third class and crew Last message received with names nine am today Please do your utmost give us this information at earliest possible moment
White Star Line

Carpathia — 16 April 1912[90]
Office of Origin: Carpathia — To: Capt Minnewaska

Only too anxious to get all names to shore Doing all possible Mr Ismay is on board Bergs and pack ice down to 41.16N and as far east as 49.30W
Rostron

Camperdown — 16 April 1912[91]
Office of Origin: Montreal — To: Marconi Station Steamer Carpathia

Please send immediately full list Titanics survivors Stop Send rush message stating whether Charles M Hays Thornton Davidson and Allison are rescued Stop Please rush messages through immediately communication Sable Island established
Arcon [Canadian Marconi Co.]

Camperdown — 16 April 1912[92]
Office of Origin: Montreal — To: Marconi Operator Carpathia

Please add Markland Molson to names Titanics passengers information particularly required about Stop Rush information regarding Chas M Hays immediately Stop Advise when we may expect full list names survivors
Canadian Marconi Co

Camperdown — 17 April 1912[93]
Office of Origin: NYK — To: Commander MPA [Carpathia]

Marconi quickly full particulars of what passengers you have from Titanic and what disposition you expect to make of them Also any particulars of Titanic her condition and whether she is in tow and destination.
Chas P Sumner
[Cunard Line]

Siasconset — 17 April 1912[94]
Office of Origin: New York — To: Operator Carpathia

Is Colonel Astor aboard ~~Give me answer Head Office quick Give this precedence~~
F.M. Sammis
[American Marconi Co.]

Camperdown
Office of Origin: New York

17 April 1912[95]
To: Commander Carpathia

It is of vital importance that we should have the names of the survivors of the Titanic that are on board your steamer Kindly wire us to New York promptly

White Star Line

Camperdown
Office of Origin: MPA [Carpathia]

17 April 1912 (7.44 am)[96]
To: Islefrank New York [White Star Line]

Deeply regret advise you Titanic sank this morning after collision iceberg resulting serious loss life Further particulars later

Bruce Ismay[97]

Camperdown
Office of Origin: MPA [Carpathia]

17 April 1912 (5.12 pm)[98]
To: Islefrank New York

Most desirable Titanic crew aboard Carpathia should be returned home earliest moment possible Suggest you hold Cedric sailing her daylight Friday unless you see any reason contrary Propose returning in her myself Please send outfit of clothes including shoes for me to Cedric Have nothing of my own Please reply

Yamsi
[Ismay]

Siasconset
Office of Origin: NYK

17 April 1912 (6.40 pm)[99]
To: Bruce Ismay "Carpathia"

So thankful you are saved but grieving with you over terrible calamity Shall sail Saturday to return with you Florence Cable ends[100] Accept my deepest sympathy Horrible catastrophe Will meet you aboard Carpathia after docking Is Widener aboard

Franklin

Siasconset
Office of Origin: Carpathia

18 April 1912 (5.21 am)[101]
To: Islefrank NYK

Please join Carpathia quarantine if possible

Yamsi

Siasconset | 18 April 1912 (5.26 am)[102]
Office of Origin: Carpathia | To: Islefrank NYK

Send responsible ship officer and fourteen white star sailors in two tug boats – to take off thirteen Titanic boats at quarantine
Yamsi

Siasconset | 18 April 1912 (5.29 am)[103]
Office of Origin: Carpathia | To: Islefrank NYK

Widener not aboard Hope see you quarantine Please cable wife am returning Cedric
Yamsi

Siasconset | 18 April 1912 (7.23 am)[104]
Office of Origin: Carpathia | To: Islefrank NYK

Very important you should hold Cedric daylight Friday for Titanic crew Reply
Yamsi

Siasconset | 18 April 1912 (7.26 am)[105]
Office of Origin: NYK | To: Ismay "Carpathia"

Have arranged forward crew Lapland sailing Saturday calling Plymouth We all consider most unwise delay Cedric considering all circumstances
Franklin

Siasconset | 18 April 1912 (7.45 am)[106]
Office of Origin: Carpathia | To: Islefrank NYK

Think most unwise keep Titanic crew until Saturday Strongly urge detain Cedric sailing her midnight if desirable
Yamsi

Siasconset | 18 April 1912 (8.19 am)[107]
Office of Origin: Carpathia | To: Islefrank NYK

Unless you have good and substantial reason for not holding Cedric please arrange do so Most undesirable have crew NYK so long
[Yamsi]

Seagate | 18 April 1912 (8.20 pm)[108]
Office of Origin: NYK | To: Rostron Commander S/S Carpathia

Message eighteenth only received 3 pm dilapsi bedding etc will be aesculiae aesalon relief for haaring expect sail Carper friday afternoon how much bothroph potentibus aestuatios for docking scenica alabado north side pier 54 Titanic marmarron membrarono tonight lower floor no omalosomes issued to herkruisen haarring prior akoluth dock petris doctor will simply put man aboard alabado and adenoideo scenica ordenadero to dock

C.P. Sumner
[Cunard Line]

MESSAGES RELATING TO THE RECOVERY OF BODIES

Camperdown | 17 April 1912[109]
Office of Origin: Halifax, N.S. | To: Mackay Bennett

Bring all bodies you may recover back to Port Do not bury any at sea Acknowledge.

Ward
[Mackay-Bennett Cable Co.]

Camperdown | 16 April 1912[110]
Office of Origin: Minia | To: Western Union

For Carlton My position too far from scene of accident to communicate with Titanic or relieving vessels My coal too low to warrant my going so far east now Dense fog but hope reach Halifax early morning Had hard time in pack ice all night last night Report today Titanic proceeding under steam Understand list survivors sent Cape Race Am unable confirm this at present Accident occurred 600 miles from Halifax Carpathia Baltic Olympic reported on scene

Decarteret

Camperdown | 16 April 1912[111]
Office of Origin: Halifax NS | To: Capt. Decarteret Cable Ship Minia

From Carlton When do you expect arrive Halifax How soon could you be ready for sea if necessary
Laidlaw

Camperdown | 16 April 1912[112]
Office of Origin: Minia | To: Laidlaw Western Union HX

Wednesday morning Fog still dense Can sail by night if coal obtainable Let my wife know Carpathia reported on way to New York
Decarteret

Camperdown | 16 April 1912[113]
Office of Origin: Minia | To: Franklyn Cunard Company on Tobin St Halifax

Please arrange to rush coal on Minia tomorrow morning as I may be required to sail at night for Titanic position.
Decarteret

Camperdown | 16 April 1912[114]
Office of Origin: Minia | To: Laidlaw Western Union HX

For Carlton If towing or assistance necessary I am better equipped with steel ropes than an ordinary vessel but no rooms for passengers.
Decarteret

Camperdown | 17 April 1912[115]
Office of Origin: Minia | To: Laidlaw Western Union

Expect reach wharf 9.45 Inform Decarteret please[116]
Decarteret

PERSONAL MESSAGES

The primary interest of the Marconi Company in having wireless equipment on board passenger liners was commercial: to send and receive chargeable personal messages on behalf of passengers. In the days leading up to the sinking, the majority of the messages dealt with by Jack Phillips and Harold Bride would have been of this type. Many passengers on board ships used wireless to communicate information about their expected arrival, the condition of the voyage, and even business negotiations with friends, family and colleagues on land. Whilst messages such as these are not directly relevant to the use of wireless during the disaster, one set is included to illustrate the contrast between the ordinary use of wireless communication on board ship, and the chaos which ensued after the sinking. These are the surviving messages between the actress Dorothy Gibson and film-maker Jules Brulatour.

The Marconi Archives contain hundreds of messages from panicking relatives, desperate to find out whether their loved ones aboard the *Titanic* had survived. The vast majority of these messages were never transmitted to the *Carpathia*. The messages on the following pages are merely a small selection, illustrating the urgent attempts to contact friends and family and, for some, the relieved congratulations sent to survivors. Although messages from the shore to the *Carpathia* were unlikely ever to reach their destination, Cottam and Bride did manage to send at least some of the messages from the survivors, and many of these were being sent right up to the moment the *Carpathia* docked in New York on the evening of 18 April. The following pages again contain only a handful of these poignant messages, often reporting the safety of some family members but the loss of others. It is perhaps here, in these short and simple messages, that the real scale of the tragedy can best be seen.

The two final messages are notable for their illustration of the resourcefulness shown in the face of tragedy. Among the survivors were Leila Meyer (née Saks) of Saks & Co. and Julia Cavendish (née Siegel) of Siegel-Cooper & Co., both large New York clothing firms. In spite of having both lost their husbands in the sinking, they appear to have immediately volunteered the resources available to them, probably to clothe the surviving officers and crew of the *Titanic*, who were required to appear before an inquiry. The two telegrams were probably sent after the *Carpathia* had docked in New York and provide an unusual insight into the practical as well as the emotional consequences of the disaster.

right
Message from Jules Brulatour to Dorothy Gibson, 12 April. (Oxford, Bodleian Library, MS. Marconi 285, fol. 1.)

below
Message from Jules Brulatour to Dorothy Gibson, 13 April. (Oxford, Bodleian Library, MS. Marconi 285, fol. 3.)

right
Message from Jules Brulatour to Dorothy Gibson, 14 April. (Oxford, Bodleian Library, MS. Marconi 285, fol. 2.)

FORM No.
Sent date
WU 13
MARCONI WIRELESS TELEGRAPH COMPANY OF AMERICA.
27 WILLIAM STREET, (Lord's Court Building), NEW YORK.
No. 3 Cape Cod STATION Apr 12 1912
Prefix S Code Words 10
Office of Origin C New York
Service Instructions:
SW

CHARGES TO PAY.		
Marconi Charge	5	00
Other Line Charge		
Delivery Charge		
TOTAL	5	00

Station sent to	Time Sent	By whom sent
MGY	10p m.	Auto

TO Dorothy Gibson Titanic SW
will do everything make you completely happy love you madly Julie

READ THE CONDITIONS PRINTED ON THE BACK OF THE FORM.

MS Marconi 285

FORM No. 1-
MARCONI WIRELE[SS]
27 WILLIAM
No. 1 Capecod STATION Apr 13 1912
Prefix S Code S Words 11
Office of Origin C New York
Service Instructions:

CHARGES TO PAY.		
Marconi Charge	5	50
Other Line Charge		
Delivery Charge		
TOTAL	5	50

Station sent to	Time Sent	By whom sent
MGY	10p m.	Auto

TO Miss D Gibson Titanic SW
It cause no happiness without mutsie never allow you leave again Julie

READ THE CONDITIONS PRINTED ON THE BACK OF THE FORM.

Sent date
[COMPA]NY OF AMERICA.
NEW YORK.
Apr 14 1912

CHARGES TO PAY.		
Marconi Charge	5	00
Other Line Charge		
Delivery Charge		
TOTAL	5	00

Station sent to	Time Sent	By whom sent
MGY		Auto

TO Miss Dorothy Gibson Titanic
Fancy adjoining room Great Northern refused Mutsie Cable Decision bleed Julie

READ THE CONDITIONS PRINTED ON THE BACK OF THE FORM.

No. 1

MARCONI WIRELESS TELEGRAPH COMPANY OF CANADA, LIMITED

MONTREAL

CAMPERDOWN

No. ... STATION ... APR 17 1912 ... 19

Prefix ... Code ... Words ... CHARGES TO PAY.

Office of Origin New York

Service Instructions: Via C.E.

Marconi Charge

Other Line Charge

Delivery Charge

Total

Station sent to | Time sent | By whom sent

NOT TRANSMITTED

READ THE CONDITIONS PRINTED ON THE BACK OF FORM.

To: Miss Dorothy Gibson Carpathia

worried to death please wire if safe and condition at once

J E Brulatour

No. 3—6-29-11—20M. F'w'ding Charge ... F'w'ded Date ...

MARCONI WIRELESS TELEGRAPH COMPANY OF AMERICA.

27 WILLIAM STREET, (Lord's Court Building), NEW YORK.

SIASCONSET

No. ... STATION APR 18 1912 ... 191

Prefix ... Code ... Words ...

Office of Origin Carpathia

Service Instructions:

Station Rec'd from	Time Rec'd	By whom Rec'd
MPA	7.21 m	G
Forwarded to	**Time F'w'd.**	**By whom F'w'd.**
NY	731 A m.	JCD

To: Jules E Brutalom 31 East 27th St Nyk

Safe picked up by Carpathia dont worry

Dorothy

above
Message from Jules Brulatour to Dorothy Gibson (not transmitted), 17 April. (Oxford, Bodleian Library, MS. Marconi 272, fol. 3.)

left
Message from Dorothy Gibson to Jules Brulatour, 18 April. (Oxford, Bodleian Library, MS. Marconi 286, fol. 1.)

PERSONAL MESSAGES: DOROTHY GIBSON AND JULES BRULATOUR

Cape Cod — 12 April 1912[117]
Office of Origin: New York — To: Dorothy Gibson Titanic

Will do everything make you completely happy Love you madly
Julie

Cape Cod — 13 April 1912[118]
Office of Origin: New York — To: Miss D Gibson Titanic

It cause no happiness without Mutsie Never allow you leave again
Julie

Cape Cod — 14 April 1912[119]
Office of Origin: New York — To: Miss Dorothy Gibson Titanic

Fancy adjoining room Great Northern Referred Mutsie Cable decision Bleed
Julie

Camperdown — 16 April 1912[120]
Office of Origin: New York — To: Dorothy Gibson Carpathia
(Non Transmitted)

Will be worried to death till I hear from you What awful agony
Julie

Camperdown — 17 April 1912[121]
Office of Origin: New York — To: Miss Dorothy Gibson Carpathia
(Non Transmitted)

Worried to death Please wire if safe and condition at once
J E Brulatour

Siasconset — 18 April 1912[122]
Office of Origin: Carpathia — To: Jules E Brulatour 31 East 27th St NYK

Safe picked up by Carpathia dont worry
Dorothy

Camperdown — 15 April 1912[123]
Office of Origin: New York — To: Isham Psgr SS Titanic

Please answer collect you are safe and likely to arrive
Ed T Isham

Camperdown — 15 April 1912[124]
Office of Origin: New York — To: Mrs J Bradley Cummins Carpathia

Mother and all well So happy you are all right
Gibly

Camperdown — 16 April 1912[125]
Office of Origin: Milwaukee Wis — To: Mrs G.G. Crosby Carpathia

Have reports you and sisters safely Is father with you Hurry answer
Fred Crosby

Camperdown — 16 April 1912[126]
Office of Origin: Montreal — To: Capt S/S Virginian

Please advise immediately if H J Allison and wife were rescued by you
Donellson

Camperdown — 16 April 1912[127]
Office of Origin: New York — To: Purser S/S Carpathia

Is Loring survivor Answer collect
Plunkett 49 Bway

Camperdown — 16 April 1912[128]
Office of Origin: New York — To: Washington Dodge Carpathia

Congratulations on escape Am exceedingly anxious know whether Edgar Meyer and wife and Isidor Strauss and wife are saved Kindly send Marconigram
C. Altschul

Camperdown
Office of Origin: Glace Bay
(London via Marconi)

16 April 1912[129]
To: Thulium Sir Cosmo Lady Duff Gordon on Carpathia

Glorious rejoicing Belief always made us know you safe Love
Eleanor

Camperdown
Office of Origin: Montreal

16 April 1912[130]
To: Capt SS Carpathia

Is H J Allison and wife and family on your ship Reply our expense
Johnstone McConnell & Allison

Camperdown
Office of Origin: New York

16 April 1912[131]
To: Nella Goldenberg Carpathia

Are you and Sam safe and well
Cozzens

Camperdown
Office of Origin: New York

16 April 1912[132]
To: Mrs W T Graham Carpathia

Waiting anxiously for message All well here
Will

Camperdown
Office of Origin: NYK

17 April 1912[133]
To: Capt of Carpathia

Is Benjamin Guggenheim aboard Please answer without delay
Benjamin Guggenheim
St Regis Hotel

Camperdown
Office of Origin: Newark NJ

17 April 1912[134]
To: Henry Stengel Carpathia

Are you both safe Answer if possible
Rothschild

Camperdown 17 April 1912[135]
Office of Origin: Montreal To: Ethel Fortune Carpathia

Sincerest sympathy Wire if I can be of any assistance
David Romsby

Camperdown 17 April 1912[136]
Office of Origin: New York To: Benjamin Guggenheim on board MPA

Anxious to hear you and Mr Straus are safe Answer
Whittenberg

Camperdown 17 April 1912[137]
Office of Origin: New York To: Col John Astor Carpathia

Hearing passengers not going Halifax Have cancelled car Love
Vincent

Camperdown 17 April 1912[138]
Office of Origin: New York To: George A Harder Carpathia

Parents anxiously waiting news Answer to me
Holmes

Camperdown 17 April 1912[139]
Office of Origin: New York To: Carpathia C'down

Have you any news of Col John A Astor
Vincent Astor

Camperdown 17 April 1912[140]
Office of Origin: New York To: Mrs Virginia Clark Carpathia

Is Walter on Carpathia We are anxious about you
Wm A Clark

Sea Gate 18 April 1912[141]
Office of Origin: [S.S.] Cedric To: Bruce Ismay S/S Carpathia Cunard Pier New York

Our thoughts with you Wish you could be with us Deepest sympathy in past and present sorrow Send you our heartiest best wishes & affection
Ivy Herman Hayes

Camperdown 19 April 1912[142]
Office of Origin: New York To: Mr and Mrs J. H. Bishop Carpathia

Congratulations and love We await you at Waldorf Astoria
Father and Mother

PERSONAL MESSAGES SENT BY SURVIVORS

Cape Sable 17 April 1912[143]
Office of Origin: MPA [Carpathia] To: Lucilation London

Sir Cosmo and Lady Duff Gordon safe
Carpathia

Camperdown 17 April 1912[144]
Office of Origin: MPA [Carpathia] To: Saks New York

Leila safe and well cared for Edgar missing

Siasconset 18 April 1912[145]
Office of Origin: Carpathia To: Percy Straus L.H. Macy and Co. Herald Square NYK

Every boat watched Father Mother not on Carpathia Hope still
Badenoch

Siasconset 18 April 1912[146]
Office of Origin: Carpathia To: Simon Madigan 4 Church St Askeaton Ireland

The ship sunk B.M. and I are safe

Siasconset 18 April 1912[147]
Office of Origin: Carpathia To: Mrs Healey 29 Stonely Drive New Brighton

Titanic gone down Am safe on Carpathia
Pattie

Siasconset | 18 April 1912[148]
Office of Origin: Carpathia | To: Bjorastrom Ruda Sweden

Safe
Hakane

Siasconset | 18 April 1912[149]
Office of Origin: Carpathia | To: Pecorini Via Po 46 Rome

Both saved
Mother

Siasconset | 18 April 1912[150]
Office of Origin: Carpathia | To: Mundy Markeaton Derby

Saved Not Tyril yet
Julia

Siasconset | 18 April 1912[151]
Office of Origin: Carpathia | To: Bowker Cottage Little Sutton Cheshire England

Safe
Ruth

Siasconset | 18 April 1912[152]
Office of Origin: Carpathia | To: Tucker Albany NY

Safe on board Carpathia Bound for NYK
Bert

Siasconset | 18 April 1912[153]
Office of Origin: Carpathia | To: Jacobsohn 34 Anson Road Cricklewood London

Don't be alarmed Sydney may be on another boat
Jacobsohn

Siasconset | 18 April 1912[154]
Office of Origin: Carpathia | To: Norris Williams St Martin Lane Phila PA

Father not seen No hope Arrive Carpathia Wednesday New York
Richard

right
Message from Leila Meyer to Saks, New York, 17 April. (Oxford, Bodleian Library, MS. Marconi 272, fol. 7.)

below
Message from Pattie to Mrs Healey, 18 April. (Oxford, Bodleian Library, MS. Marconi 286, fol. 2.)

No. 1a.

MARCONI WIRELESS TELEGRAPH COMPANY OF CANADA, Limited

MONTREAL

No. 22 Camperdown *STATION* Apr 17 *19* 12

Prefix S *Code* *Words* 8

Office of Origin

Service Instructions: MPA.

via W.H. via MCA.

Station Rec'd from	Time Rec'd	By whom Rec'd
SB.	300 pm.	D
Stations sent to	**Time Sent**	**By whom sent**
37 HH	308 pm.	M H

To Saks Newyork

Leila safe and well cared for edgar missing

No. 3—6-29-11—20M F'w'ding Charge 25+3-80=4.05 F'w'ded Date APR 18 1912

MARCONI WIRELESS TELEGRAPH COMPANY OF AMERICA.

27 WILLIAM STREET (Lord's Court Building), NEW YORK.

No. 18 SIASCONSET *STATION* APR 18 1912 *191*

Prefix *Code* *Words* 14

Office of Origin Carpathia

Service Instructions: SIASCONSET

Station Rec'd from	Time Rec'd	By whom Rec'd
MPa	6.12 am.	S
Forwarded to	**Time F'w'd.**	**By whom F'w'd.**
MY	6/6 am.	JC X

To: Mrs Healey 29 Stoneby Drive New Brighton.

Titanic gone down am safe on Carpathia

Pattie

Siasconset 18 April 1912[155]
Office of Origin: Carpathia To: Cook William Jeweller College Liberty Missouri

Safe Carpathia
Collett

Sea Gate 18 April 1912[156]
Office of Origin: MPA [Carpathia] To: Mrs H S Force 18 East 37 St New York

Mrs Astor safe

Sea Gate 18 April 1912[157]
Office of Origin: Carpathia To: Hamilton 32 Bedford Place London

Saved Cable Dick Lost address
Fanny

Sea Gate 18 April 1912[158]
Office of Origin: MPA [Carpathia] To: Pears Isleworth England

Edith safe All hope for Tom

Sea Gate 18 April 1912[159]
Office of Origin: Carpathia To: White Side and Blank Newark NJ

Safe on Steamer Carpathia Let nobody worry Bound for NY Arrive Thursday
Blank

Sea Gate 18 April 1912[160]
Office of Origin: MPA Carpathia To: Saks and Co. NY

36 mens medium flannel shirts 12 mens ditto drawers 12 pairs socks Deliver immediately at pier 54 to officer C.H. Lightoller.
Leila

Sea Gate 18 April 1912[161]
Office of Origin: Carpathia To: Siegel Cooper and Co. Sixth Avenue Nyk

25 coats 19 trousers medium weight for destitute Deliver immediately at Pier 54 to officer C.H. Lightoller
Julia

GLOSSARY

CALLSIGNS

Each ship and shore station was identified by a three-letter callsign. In the transcripts, the names of the stations have been added in square brackets for clarity, where they can be identified, but a list of all callsigns which appear in the transcripts is given below.

SHIPS

DDR	SS *Amerika*
MJL	SS *Antillian*
MKL	SS *Asian*
MTI	SS *Athinai*
MBC	SS *Baltic*
DKB	SS *Berlin*
SBA	SS *Birma*
DDB	SS *Blucher*
MEL	SS *Bohemian*
MLB	SS *La Bretagne*
MWL	SS *Californian*
MAA	RMS *Carmania*
MRA	RMS *Caronia*
MPA	RMS *Carpathia*
MLC	RMS *Celtic*
NDG	USS *Chester*
DDC	SS *Cincinnati*
MCN	SS *Corsican*
MBT	SS *Elizabethville*
MEA	RMS *Franconia*
DFT	SS *Frankfurt*
DKN	SS *George Washington*

DHO	SS *Hellig Olav*
MHS	SS *Hudson*
MEI	SS *Iroquois*
DKP	SS *Kronprinz Wilhelm*
MAM	SS *Luisiana*
MDT	SS *Macedonia*
MMB	CS *Mackay-Bennett*
UMO	SS *Marengo*
MGA	RMS *Mauretania*
MZC	SS *Megantic*
MMV	SS *Mesaba*
ANM	CS *Minia*
MMW	SS *Minnewaska*
MLQ	SS *Mount Temple*
MEN	SS *Navahoe*
MHA	SS *Noordam*
NMN	USS *North Carolina*
MVO	SS *Oceania*
MKC	RMS *Olympic*
MJB	SS *Oravia*
MZN	SS *Parisian*
DDF	SS *Pisa*
DDS	SS *President Grant*
MPK	SS *Princess Adelaide*
DDZ	SS *Prinz Adalbert*
DKF	SS *Prinz Friedrich Wilhelm*
DPS	SS *Prinz Sigismund*
MLP	SS *La Provence*
MER	SS *Royal Edward*
NRZ	USS *Salem*
MLS	SS *La Savoie*
MNC	SS *Scandinavian*
MGY	RMS *Titanic*
MLT	SS *La Touraine*
DUS	SS *United States*
MGN	SS *Virginian*
DYA	SS *Ypiranga*

SHORE STATIONS

NAD	Boston, Massachusetts, Navy yard
NAH	Brooklyn, New York, Navy yard
MHX	Camperdown, Halifax, Nova Scotia
NAE	Cape Cod, Highland Light, Massachusetts, Naval station
MCC	Cape Cod, Massachusetts
MCE	Cape Race, Newfoundland
MSB	Cape Sable, Nova Scotia
WQ	Eastport, Maine
WS	New London, Connecticut
NMK	New York, Naval station
NAF	Newport, Rhode Island, Torpedo station
PJ	Point Judith, Rhode Island
ZZ	Poldhu, England
NAC	Portsmouth, New Hampshire, Navy yard
MSD	Sable Island, Nova Scotia
MSK	Sagaponack, New York
MSC	Siasconset, Massachusetts
MSE	Sea Gate, New York
MSJ	St John, Partridge Island, New Brunswick

RADIO CODES

Wireless operators used a number of codes and abbreviations, both official and unofficial, to enable more concise and rapid communication. Many of these can be seen throughout the transcripts. Although the meaning of some is evident, others require some explanation. A selection of codes and their meanings is given below. For this glossary, most codes have been capitalised, but they can be seen in both upper and lower case in the transcripts, depending on the habit of the operator.

BI	Waiting
CQ	'Calling all ships'
CQD	'Distress – calling all ships'
DDD	'Silence' / 'Shut up'
DE	'From' / 'This is'
G	'Go ahead' / 'Start sending'
K	'Please reply'
Nil	No messages exchanged
NR	No response
OM	'Old man' (friendly form of address used between operators)
P	Passenger message (commercial)
RD	'Received'
S	Service message (not commercial traffic)
SG / MSG	Prefix to Master's Service Message (captain must acknowledge the message)
SOS	'Distress' (introduced in 1908)
STD BI	'Please wait' / 'Stand by'
TIS	'End of message'
TR	Time rush (exchange of signals and current time)
V	Letter V commonly sent when testing apparatus
Xs	Atmospherics which interfere with the signal

NOTES

INTRODUCTION

1. Calchas, 'The Operating Side of Wireless Telegraphy as a Career', *The Marconigraph*, vol. I, 1911–12, pp. 8–12.
2. MS. Marconi 255, fols. 10–16: copy agreement between the Marconi International Marine Communication Company and the Oceanic Steam Navigation Company, 9 August 1909 (with handwritten additions referring to letters dated 28 April and 3 May 1911).
3. 'An Epic Tragedy of the Sea: Wireless – the Wonder Worker', *The Marconigraph*, vol. II, 1912–13, pp. 36–40.
4. Calchas, 'The Operating Side of Wireless Telegraphy as a Career'.
5. MS. Marconi 2040: staff salary book 1902–12, p. 44.
6. Leafield, near Oxford, was one of several wireless stations involved in the interception of enemy signals for the Admiralty Intelligence Division during World War 1. See MS. Marconi 335: correspondence with the Admiralty, 1912–19.
7. MS. Marconi 2040: staff salary book 1902–12, p. 169.
8. Ibid. p. 90.
9. MarconiCalling website: www.marconicalling.co.uk/museum/html/events/events-i=51-s=1.html, accessed 11 August 2011.
10. MS. Marconi 264, fol. 33: message *Noordam* to *Titanic* via *Caronia*, received 2.31 pm, sent 2.45 pm 14 April.
11. MS. Marconi 264, fol. 14: message *Caronia* to *Californian*, 5.08 pm 15 April.
12. MS. Marconi 261, fols. 141–6: statement of Stanley Adames, *Mesaba* operator, [May] 1912.
13. MS. Marconi 285, fol. 1: message Jules Brulatour to Dorothy Gibson, 10 pm 12 April.
14. MS. Marconi 264, fol. 18: message Titanic to CQD [picked up by *La Provence*] [no time] 14 April.
15. MS. Marconi 264, fol. 73: message Titanic to CQD [picked up by *Celtic*] 5.21 am 14 April.
16. W.J. Baker, *A History of the Marconi Company* (London: Methuen, 1970), p. 139.
17. MS. Marconi 264, fol. 61: message *Californian* to *Virginian*, 3.55 am 15 April.
18. MS. Marconi 264, fol. 63: message *Virginian* to *Californian*, 4.00 am, 15 April.
19. MS. Marconi 264, fol.55: message *Olympic* to *Titanic*, 11.50 pm 14 April.
20. MS. Marconi 261, fol. 309: *Virginian* PV.
21. MS. Marconi 261, fol. 310: *Virginian* PV.
22. MS. Marconi 258, fols. 163–5: printed letter from the Marine Company to Sir R. Ellis Cunliffe with extracts from reports of steamers, 1 May 1912.
23. MS. Marconi 264, fol. 94: message *Olympic* to *Carpathia*, 2.35 pm 15 April.
24. MS. Marconi 264, fols. 130–32: message *Carpathia* to *Olympic*, [no time] 15 April.
25. MS. Marconi 285, fols. 9, 11–13 (messages handled by Cape Sable 6.51–7.10 pm 17 April); 107 (handled by Sagaponack 1.00 pm 18 April); MS. Marconi 286, fols. 4, 42 (handled by Siasconset 7.56 am and 8.14 am 18 April).
26. MS, Marconi 286, fol. 1: message Gibson to Brulatour 7.27 pm 18 April (sent via Siasconset).

27. MS. Marconi 261, fol. 196: message New York to *Minnewaska*, 7.05 am 16 April.
28. MS. Marconi 261, fol. 200: message *Carpathia* to *Minnewaska*, 1.00 pm 16 April.
29. MS. Marconi 261, fols. 181–2: *Minnewaska* PV, 16 April.
30. MS. Marconi 256, fol. 45: letter Cottam to Turnbull, n.d. [April/May 1912].
31. MS. Marconi 285, fol. 86: message Cottam to Ingram MI 12.55 am 18 April.
32. MS. Marconi 266, fol. 121: message Ismay to Islefrank [White Star Line], New York, 7.44 am 17 April.
33. MS. Marconi 286, fols. 61, 63: messages from Leila [Meyer] and Julia [Siegel], 18 April.
34. MS. Marconi 256, fols. 5–13: messages between Marine Company and Camperdown station, 16–17 April.
35. MS. Marconi 256, fols. 16–22: messages about attempts to obtain information for Southampton, 18 April.
36. Ibid.
37. MS. Marconi 256, fols. 27–41: messages and correspondence with and about the Newfoundland Governor seeking news 16–30 April.
38. MS. Marconi 260, fol. 61: *Asian* PV, 14 April.
39. MS. Marconi 266, fol. 26: message (not transmitted) Franklin, [White Star Line] to Captain Haddock, *Olympic*, 15 April.
40. MS. Marconi 260, fols. 85–6: Printed report on *Asian*'s assistance to German Tank Steamer *Deutschland*.
41. MS. Marconi 279, fol. 10: message *New York Herald* to Jacques Futrell, 7.54 pm 15 April.
42. Ibid. fol. 23: message *New York Herald* to wireless operator, *Olympic*, 15 April.
43. MS. Marconi 280, fol. 5: message (not transmitted) *New York Times* to Ismay, *Carpathia*, 17 April.
44. MS. Marconi 281, fol. 86: message (not transmitted) *Montreal Star* to W.T. Stead, 17 April.
45. Ibid. fol. 99: message (not transmitted) *New York Evening Journal* to Captain Carpathia, 17 April.
46. MS. Marconi 1489–91.
47. MS. Marconi 257, fols. 3–4: confirmation of telegram from Marine Company to American Company, 18 April 1912.
48. MS. Marconi 258, fols. 163–5: printed letter from the Marine Company to Sir R. Ellis Cunliffe with extracts from reports of steamers, 1 May 1912.
49. MS. Marconi 258, fols. 166–8: printed letter from the Marine Company to Sir R. Ellis Cunliffe with copy of procès-verbal of the SS *Carpathia* and copies of service form messages relating to ice, 7 May 1912.
50. MS. Marconi 258, fols. 149–56: draft statement of G.E. Turnbull, n.d. [May 1912].
51. Ibid.
52. MS. Marconi 258, fols. 157–62: letter from [G.E. Turnbull] to Guglielmo Marconi, 13 June 1912.
53. MS. Marconi 262, fol. 23: TS summary of Cape Race PV, 14 April.
54. MS. Marconi 291–2: Minutes of Evidence of the British Inquiry 1912, p. 202.
55. See John Booth and Sean Coughlan: *Signals of Disaster* (Westbury: White Star, 1993).
56. MS. Marconi 258, fols 172–89: Substitute for the procès-verbaux of the *Titanic* from 12th April ... also for the *Carpathia*, from the 14th to 18th April 1912.
57. MS. Marconi 258, fols. 190–201: copies reproduced from the original documents of Masters' Service Messages received by the *Titanic* between 11th and 14th April 1912.
58. MS. Marconi 258, fols. 102–9: Marine Company to Cunliffe, 30 May 1912.
59. MSS. Marconi 287–9: transcripts of proceedings of the US investigation, pp. 215–53.
60. Ibid., pp. 348–72.
61. Ibid., pp. 3430–37.
62. MS. Marconi 290: Report of US Inquiry, 1912.
63. MSS. Marconi 291–2: Minutes of Evidence of the British Inquiry, 1912.
64. MS. Marconi 293: Report of the British Inquiry, 1912.
65. MS. Marconi 295: Convention for the Safety of Life at Sea, 1914, Articles 31, 34 and 35.

66. Baker, p. 142.
67. MS. Marconi 256, fol. 66: telegram G.A. Phillips to Marine Company, 15 April.
68. MS. Marconi 256, fol. 67: confirmation of telegram Marine Company to G.A. Phillips, 15 April.
69. MS. Marconi 256, fols. 66–147: correspondence about the loss of Phillips, condolences and appreciation of the role of Phillips and Bride, April–November 1912.

PRIMARY SOURCES

1. In a statement submitted to the US Inquiry and to the Marconi Company, Bride stated that he had prepared two copies of the PV but that he and Phillips left too hurriedly to take them (US Inquiry, Day 14, testimony of Harold Bride, www.titanicinquiry.org; accessed 15 September 2011).
2. MS. Marconi 287, fol. 75; MS. Marconi 289, fols. 184 and 188.
3. MS. Marconi 287, fols. 283–9.
4. MS. Marconi 261, fols. 208–10.
5. MS. Marconi 261, fols. 308–10.
6. MS. Marconi 261, fols. 230–36.
7. MS. Marconi 261, fol. 243.
8. MS. Marconi 260, fols. 220–2. This was prepared, from other ships' PVs and the wireless operator's memory, for the US Inquiry.
9. MS. Marconi 261, fol. 80; translated from the original German.
10. MS. Marconi 260, fols. 120–21.
11. MS. Marconi 260, fols. 206–11.
12. MS. Marconi 260, fol. 231.
13. MS. Marconi 260, fols. 62–3.
14. MS. Marconi 260, fols. 81–4.
15. MS. Marconi 261, fols. 319–21; translated from the original German.
16. MS. Marconi 261, fol. 249.
17. MS. Marconi 261, fols. 194–200.
18. MS. Marconi 262, fols. 46–9.
19. MS. Marconi 262, fols. 75–86.
20. *New York Times*, 19 April 1912.
21. MS. Marconi 263, fol. 116.
22. MS. Marconi 263, fol. 67.
23. MS. Marconi 263, fol. 69.
24. MS. Marconi 263, fol. 137.
25. MS. Marconi 263, fol. 71.
26. MS. Marconi 263, fol. 18.
27. MS. Marconi 263, fol. 73.
28. MS. Marconi 263, fol. 87.
29. MS. Marconi 263, fol. 65.
30. MS. Marconi 263, fol. 112.
31. MS. Marconi 263, fol. 135.
32. Note by operator or captain of *Baltic*'s distance (in miles) from *Titanic*'s position.
33. MS. Marconi 263, fol. 29.
34. This appears to be a summary of several messages written by the operator for the captain.
35. MS. Marconi 263, fol. 27.
36. MS. Marconi 263, fol. 55.
37. MS. Marconi 263, fol. 53.
38. MS. Marconi 263, fol. 63.
39. MS. Marconi 263, fol. 61.
40. In order to justify a change of course, the captain would require an official Masters' Service Message (with the prefix 'SG').
41. MS. Marconi 287, fol. 59: transcript of the proceedings of the US Inquiry.
42. See testimonies of Harold Bride and the *Titanic*'s officers in MSS. Marconi 287–9.
43. Ibid. (see especially testimony of Harold Bride: MS. Marconi 287, fols. 316–27).
44. US Inquiry, Day 8, testimony of Cyril Evans, www.titanicinquiry.org; accessed 15 September 2011.
45. For advances in this field, see the Convention for the Safety of Life at Sea 1914, Articles 8–10 (MS. Marconi 295).
46. MS. Marconi 263, fol. 7.
47. MS. Marconi 263, fol. 5.
48. MS. Marconi 263, fols. 22–3.
49. MS. Marconi 263, fol. 31.
50. MS. Marconi 263, fol. 35.
51. MS. Marconi 263, fol. 49.
52. This section is a note by Stanley Adames, wireless operator on the *Mesaba*, recording what appears to be the *Titanic*'s reply. This was simply an

immediate reply from the operator on the *Titanic* and not an official response from the captain. This message does not appear to have been passed to the bridge.

53. MS. Marconi 263, fol. 3.
54. MS. Marconi 263, fol. 9.
55. MS. Marconi 263, fols. 11–12.
56. MS. Marconi 263, fol. 16.
57. This message was also sent to the *Californian*.
58. MS. Marconi 263, fol. 20.
59. MS. Marconi 266, fol. 40.
60. MS. Marconi 263, fol. 57.
61. MS. Marconi 263, fol. 59.
62. MS. Marconi 263, fol. 25.
63. MS. Marconi 263, fol. 47.
64. MS. Marconi 263, fol. 37.
65. MS. Marconi 263, fol. 33.
66. Wireless operator's note.
67. MS. Marconi 263, fol. 41.
68. MS. Marconi 263, fol. 43.
69. MS. Marconi 263, fol. 45.
70. MS. Marconi 263, fol. 75.
71. MS. Marconi 263, fol. 81.
72. US Inquiry, Day 3, testimony of Philip Franklin, www.titanicinquiry.org, accessed 15 September 2011.
73. MS. Marconi 287, fol. 75.
74. US Inquiry, Day 5, testimony of C.H. Lightoller, www.titanicinquiry.org, accessed 15 September 2011.
75. US Inquiry, Day 11, testimony of J. Bruce Ismay, www.titanicinquiry.org, accessed 15 Septemeber 2011.
76. MS. Marconi 266, fol. 97.
77. MS. Marconi 263, fols. 118–19.
78. MS. Marconi 266, fol. 4.
79. MS. Marconi 266, fol. 1.
80. MS. Marconi 266, fol. 3.
81. MS. Marconi 266, fol. 28.
82. MS. Marconi 266, fol. 26.
83. MS. Marconi 266, fol. 33.
84. MS. Marconi 263, fol. 93.
85. MS. Marconi 265, fols. 1–2.
86. MS. Marconi 263, fols. 98–100.
87. MS. Marconi 263, fols. 131–3.
88. MS. Marconi 266, fol. 147.
89. MS. Marconi 266, fol. 130.
90. MS. Marconi 263, fol. 153.
91. MS. Marconi 266, fol. 48.
92. MS. Marconi 266, fol. 51.
93. MS. Marconi 266, fol. 163.
94. MS. Marconi 286, fol. 51.
95. MS. Marconi 266, fol. 156.
96. MS. Marconi 266, fol. 121.
97. This message was the subject of much controversy owing to the fact that it was not sent until almost forty-eight hours after it was written, in spite of its evident importance.
98. MS. Marconi 266, fols. 122–3.
99. MS. Marconi 286, fols. 59–60.
100. The first section of this message is a telegram from Ismay's wife, Florence, quoted by Franklin in his message.
101. MS. Marconi 286, fol. 53.
102. MS. Marconi 286, fol. 52.
103. MS. Marconi 286, fol. 54.
104. MS. Marconi 286, fol. 56.
105. MS. Marconi 286, fol. 55.
106. MS. Marconi 286, fol. 57.
107. MS. Marconi 286, fol. 58.
108. MS. Marconi 286, fols. 67–9.
109. MS. Marconi 266, fol. 61.
110. MS. Marconi 266, fols. 67–8.
111. MS. Marconi 266, fol. 89.
112. MS. Marconi 266, fol. 80.
113. MS. Marconi 266, fol. 84.
114. MS. Marconi 266, fol. 86.
115. MS. Marconi 266, fol. 91.
116. This should presumably read 'Inform Carlton please' but the mistake is replicated on every copy of the message.
117. MS. Marconi 285, fol. 1.
118. MS. Marconi 285, fol. 3.
119. MS. Marconi 285, fol. 2.
120. MS. Marconi 272, fol. 2.
121. MS. Marconi 272, fol. 4.

122. MS. Marconi 286, fol. 1.
123. MS. Marconi 278, fol. 75.
124. MS. Marconi 278, fol. 80.
125. MS. Marconi 278, fol. 52.
126. MS. Marconi 278, fol. 54.
127. MS. Marconi 278, fol. 3.
128. MS. Marconi 278, fol. 49.
129. MS. Marconi 278, fol. 41.
130. MS. Marconi 278, fol. 35.
131. MS. Marconi 278, fol. 40.
132. MS. Marconi 278, fol. 120.
133. MS. Marconi 278, fol. 67.
134. MS. Marconi 278, fol. 11.
135. MS. Marconi 278, fol. 17.
136. MS. Marconi 278, fol. 24.
137. MS. Marconi 278, fol. 25.
138. MS. Marconi 278, fol. 30.
139. MS. Marconi 278, fol. 33.
140. MS. Marconi 278, fol. 90.
141. MS. Marconi 286, fol. 62.
142. MS. Marconi 278, fol. 5.
143. MS. Marconi 285, fol. 1.
144. MS. Marconi 272, fol. 7.
145. MS. Marconi 286, fol. 3.
146. MS. Marconi 286, fol. 42.
147. MS. Marconi 286, fol. 2.
148. MS. Marconi 286, fol. 8.
149. MS. Marconi 286, fol. 13.
150. MS. Marconi 286, fol. 30.
151. MS. Marconi 286, fol. 16.
152. MS. Marconi 286, fol. 21.
153. MS. Marconi 286, fol. 10.
154. MS. Marconi 286, fol. 4.
155. MS. Marconi 286, fol. 7.
156. MS. Marconi 286, fol. 73.
157. MS. Marconi 286, fol. 76.
158. MS. Marconi 286, fol. 86.
159. MS. Marconi 286, fol. 95.
160. MS. Marconi 286, fol. 61.
161. MS. Marconi 286, fol. 63.

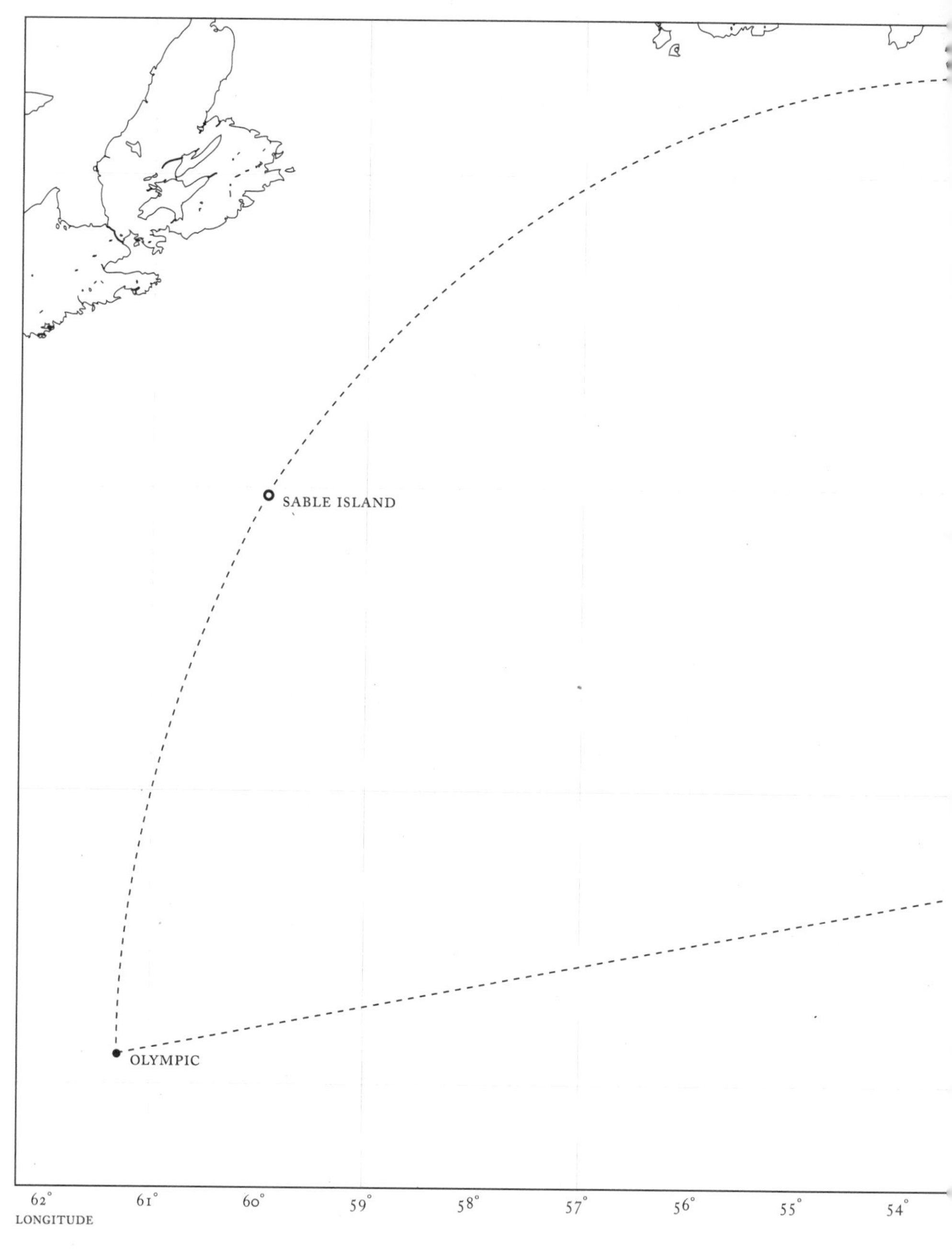

A REPRESENTATION OF THE FLOW OF MESSAGES ON THE NIGHT OF THE SINKING, FROM THE TIME OF THE FIRST DISTRESS CALL UNTIL THE CARPATHIA LEFT THE SCENE OF THE DISASTER